The Sacred Depths of Nature

Ursula Goodenough

New York   Oxford
Oxford University Press
1998

Oxford University Press

Oxford New York
Athens Auckland Bangkok Bogotá Buenos Aires Calcutta
Cape Town Chennai Dar es Salaam Delhi Florence Hong Kong Istanbul
Karachi Kuala Lumpur Madrid Melbourne Mexico City Mumbai
Nairobi Paris São Paulo Singapore Taipei Tokyo Toronto Warsaw

and associated companies in
Berlin Ibadan

Published by Oxford University Press, Inc.
198 Madison Avenue, New York, New York 10016

Oxford is a registered trademark of Oxford University Press

Library of Congress Cataloging-in-Publication Data
Goodenough, Ursula.
The sacred depths of nature / by Ursula Goodenough.
p. cm. Includes bibliographical references and index.
ISBN 0-19-512613-0
1. Biology—Philosophy. 2. Biology—Religious aspects.
3. Naturalism—Religious aspects. 4. Nature—Religious aspects.
I. Title.
QH331.G624 1998
574'.01—DC21 98-6579

Permissions and sources for art and poetry can be found in the *Notes and Further
Reading* section.

9 8 7 6 5 4 3 2 1
Printed in the United States of America
on acid free paper

for
Edna Jackson Wiltshire
Evelyn Wiltshire Goodenough Pitcher
Ursula Goodenough Stewart

Contents

Personal

No question about it: I'm writing this book because of my father. He started out as a Methodist preacher but became absorbed—no, obsessed—with a need to understand why people are religious. As Professor of the History of Religion, he poured out book after book on the ancient Jews and early Christians: their art, their texts, their motivations. And then he brought it all home, to me sitting there after dessert trying to look inconspicuous while he and the other Yale scholars drank a great deal of wine and held forth on Plato and Paul and Freud and Sartre. Dad began his famous undergraduate course, The Psychology of Religion, by announcing "I do not believe in God." He ended one of his last books by admitting "I still pray devoutly, and when I do I forget my qualifications and quibbles and call upon Jesus—and he comes to me." He was a larger-than-life father, passionate and outrageous and adored. When he died of cancer when I was twenty-two, it was almost more than any of us could bear.

I went to college with 50s expectations: find a husband, raise two children, and continue to read novels. But everything changed when I took Zoology 1 as a distribution requirement. Nothing in my girls'-school training had led me to understand that creatures are made up of cells and genes and enzymes, that life evolves, that kidneys control blood electrolytes. I was astonished. Better still, I was good at it. And Dad was quite as excited about my unexpected calling as I was. "Ursula a scientist! How splendid!" What a father.

For the next twenty-five years or so I played it straight: biology professorships, research projects, federal grants, undergraduate teaching. I still do all those things, and with as much pleasure and satisfaction as ever. But as my five children grew and there was more time for myself, my father's question returned. Why are people religious? And then: Why am I *not* religious?

But was that true? What *is* being religious anyhow? What about the way I feel when I think about how cells work or creatures evolve? Doesn't that feel the same as when I'm listening to the St.

Matthew Passion or standing in the nave of the Notre Dame Cathedral?

So I joined Trinity Presbyterian Church, and have spent the past twelve years singing in the choir, reciting the liturgy and prayers, hearing the sermons, participating in the ritual. I came to understand how this tradition, as played out in a middle-class mostly white congregation, is able to elicit states of serious reflection, reverence, gratitude, and penance. But all of it was happening in the context of ancient premises and a deep belief in the supernatural. What about the natural? Was it possible to feel such religious emotions in the context of a fully modern, up-to-the-minute understanding of Nature?

And so I started reading and talking and listening and reflecting, and out of it has emerged this book. Certainly the most important dialogue has been with Loyal Rue, who has explained to me most of what I understand about theology and philosophy and who has insisted that we scientists speak of what we know and feel. Early on I happened onto an improbable collection of people composing the Institute on Religion in an Age of Science, and while the support and input of everyone in IRAS has been seminal, this is particularly true for Gene d'Aquili, Connie Barlow, Michael Cavanaugh, Tom Gilbert, Ward Goodenough, Phil Hefner, Bill Irons, Sol Katz, Ted Laurenson, Karl Peters, Bob Schaible, and Barbara Whitaker-Johns. Kirk Jensen of Oxford University Press has provided generous and unwavering support; Carl Smith has helped me understand and experience the religious impulse; John Heuser has continuously infused his perspective and wisdom; Sine Berhanu and Jeanne Heuser have nurtured my spirituality; Pam Burroughs, Elizabeth Marincola, Sharon Olds, and Betsy Weinstock have nurtured my courage; my children—Jason, Mathea, Jessica, Thomas, and James—bless my life in countless ways; and no one can emerge from a consideration of religion without thanking William James.

Introduction

When people talk about religion, most soon mention the major religious traditions of our times. But
then, thinking further, most mention as well the
religions of indigenous peoples and of such vanished civilizations as ancient Greece and Egypt and
Persia. That is, we have come to understand that
there are—and have been—many different religions; anthropologists estimate the total in the
thousands. They also estimate that there have been
thousands of human cultures, which is to say that
the making of a culture and the making of its religion go together: Every religion is embedded in its
cultural history. True, certain religions have
attempted, and variously succeeded, in crossing cultural boundaries and "converting the heathens,"
but even here the invaded cultures put their unmis-
takable stamp on what they import, as evinced by

the pulsating percussive Catholic masses sung in Africa.

In the end, each of these religions addresses two fundamental human concerns: How Things Are and Which Things Matter. How Things Are becomes formulated as a Cosmology or Cosmos: How the universe came to be, how humans came to be, what happens after we die, the origins of evil and tragedy and natural disaster. Which Things Matter becomes codified as a Morality or Ethos: the Judaic Ten Commandments, the Christian Sermon on the Mount, the Five Pillars of Islam, the Buddhist Vinaya, the Confucian Five Relations. The role of religion is to integrate the Cosmology and the Morality, to render the cosmological narrative so rich and compelling that it elicits our allegiance and our commitment to its emergent moral understandings. As each culture evolves, a unique Cosmos and Ethos appear in its co-evolving religion. For billions of us, back to the first humans, the stories, ceremonies, and art associated with our religions-of-origin are central to our matrix.

I stand in awe of these religions. I am deeply enmeshed in one of them myself. I have no need to take on the contradictions or immiscibilities between them, any more than I would quarrel with the fact that Scottish bagpipes coexist with

Japanese tea ceremonies. And indeed, the resounding failure of Soviet Marxism to obliterate Russian Orthodoxy, and of Maoism to obliterate Buddhism, Confucianism, or Taoism, reminds us that any project designed to overthrow established cultural traditions is inherently doomed.

My concern is very different. As I witness contempory efforts to generate global understanding, I see some high-minded and idealistic people attempting to operate within an amalgam of economic, military, and political arrangements, and I find myself crying out "But wait! Where is the religion? What is really orienting this project besides fear and greed? Where is the shared cosmology and the shared morality?"

That we need a planetary ethic is so obvious that I need but list a few key words: climate, ethnic cleansing, fossil fuels, habitat preservation, human rights, hunger, infectious disease, nuclear weapons, oceans, ozone layer, pollution, population. Our global conversations on these topics are, by definition, cacophonies of national, cultural, and religious self-interest. Without a common religious orientation, we basically don't know where to begin, nor do we know what to say or how to listen, nor are we motivated to respond.

My agenda for this book is to outline the foundations for such a planetary ethic, an ethic that would make no claim to supplant existing traditions but would seek to coexist with them, informing our global concerns while we continue to orient our daily lives in our cultural and religious contexts.

Any global tradition needs to begin with a shared worldview—a culture-independent, globally accepted consensus as to how things are. From my perspective, this part is easy. How things are is, well, *how things are:* our scientific account of Nature, an account that can be called The Epic of Evolution. The Big Bang, the formation of stars and planets, the origin and evolution of life on this planet, the advent of human consciousness and the resultant evolution of cultures—this is the story, the one story, that has the potential to unite us, because it happens to be true.

But that potential can be realized under only one condition. A cosmology works as a religious cosmology only if it resonates, only if it makes the listener feel religious. To be sure, the *beauty* of Nature—sunsets, woodlands, fireflies—has elicited religious emotions throughout the ages. We are moved to awe and wonder at the grandeur, the poetry, the richness of natural beauty; it fills us with joy and thanksgiving. Our response to accounts of the *workings* of Nature, on the other hand, is decidedly less positive. The scientific version of how things are, and how they came to be, is much more likely, at first encounter, to elicit alienation, anomie, and nihilism, responses that offer little promise for motivating our allegiance or moral orientation.

It is therefore the goal of this book to present an accessible account of our scientific understanding of Nature and then suggest ways that this account can call forth appealing and abiding religious responses—an approach that can be called religious naturalism. If religious emotions can be elicited by natural reality—and I believe that they can—then the story of Nature has the potential to serve as the cosmos for the global ethos that we need to articulate. I will not presume to suggest what this ethos might look like. Its articulation must be a global project. But I am convinced that the project can be undertaken only if we all experience a solemn gratitude that we exist at all, share a reverence for how life works, and acknowledge a deep and complex imperative that life continue.

A key component of any religious cosmology is its human focus. Even as we now understand that our advent on the planet was but a moment ago, even as we now gaze into the heavens with new and urgent questions about our significance, the significance and future of humanity remain central to our religious concerns.

Religious naturalism has no problem here. Being at home with our natural selves is the prelude to ecology, both environmental and cultural, and there are many ways to see human beings as noble and distinctive even as we are inexorably part of the whole. A global ethic must be anchored both in an understanding of human nature and an understanding of the rest of Nature. This, I believe, can be achieved if we start out with the same perspective on how Nature is put together, and how human nature flows forth from whence we came.

How This Book Is Put Together

A Lutheran friend who read an early draft of this book remarked that it was set up like a Daily Devotional booklet. A Daily Devotional, he explained, contains a collection of short stories, each story followed by a religious meditation on the story's theme. Not being Lutheran, I wasn't familiar with the genre, but that's basically how this book is constructed.

The text is divided into twelve chapters. Each begins with a short story about the dynamics of Nature. Most of these stories are about biology, since this is what I best understand, and most are about biology at the level of molecules and genes and cells, since this is what cries out to be understood. The stories walk through the Epic of Evolution: the origins of the universe and the planet; the origins of chemistry and life; the workings of cells and organisms; the patterns of biological evolution and the resultant biodiversity; awareness and emotion; sex and sexuality; multicellularity and death; and speciation. Throughout, I have done my best to bridge the two cultures. For readers not versed in scientific concepts and terminology, I have made every effort to render the accounts under-standable, accurate, and meaningful. Those who know the terrain will, I hope, find themselves engaged by the analogies and narratives that are used to explain the familiar.

Then, at the end of each story, I offer a short religious response, the analogue of the Lutheran meditation. In some cases these responses draw on traditional religious concepts, most often from the Judeo-Christian tradition since that is what is most familiar to me. But for the most part each response is personal, describing the particular religious emo-tion or mental state that is elicited in me when I think about a particular facet of the evolutionary story. For example, the evolution of the cosmos invokes in me a sense of mystery; the increase in biodiversity invokes the response of humility; and an understanding of the evolution of death offers me helpful ways to think about my own death. If religious naturalism is to flourish, it will be because others find themselves called to reflect on the dynamics of Nature from their own cognitive, experiential, and religious perspectives—in which case this book will become one of an emergent series of Daily Devotionals.

Human memory, they say, is like a coat closet: The most enduring outcome of a formal education is that it creates rows of coat hooks so that later on, when you come upon a new piece of information, you have a hook to hang it on. Without a hook, the new information falls on the floor. Some readers with scanty scientific backgrounds have told me that at the time they were reading one of my stories about Nature, they felt like they understood everything I said, but the next day they couldn't remember a thing about it. No hooks, I explain. Then I remind them that there isn't going to be a test, and that as they were reading the story they were in fact creating hooks for their next encounters with scientific explanation. And then, the most important part: The point of hearing a story for the first time is not to remember it but to experience it.

THE SACRED DEPTHS OF NATURE

1

Origins of the Earth

INFINITIES AND INFINITESIMALS

Everything in our universe, including the Earth and its living creatures, obeys the laws of physics, laws that became manifest in the first moments of time. Much of what we know to be true about the physical universe, like the curvature of spacetime and the fact that electrons are both particles and waves, is very difficult to visualize, even for people who spend their lives thinking about such topics. Moreover, as physicists and mathemeticians probe ever more deeply, they present us with ever more mind-boggling concepts, like the idea that subatomic particles may in fact be minute, vibrating "superstrings" of space, that our four-dimensional universe may actually be ten-dimensional, that the observable universe may be much smaller than the true universe, and that there may be many other universes besides our own.

Fascinating as these known and speculative manifestations of physics may be, they prove not to be central to our story of life. Why? Because when Earth life was coming into being, some ten billion years after the universe had come into being, the laws of physics were a *given*. Life had no choice but to evolve in the context of quantum indeterminacy and gravitational fields and quarks held together by gluons. Therefore, while these facts underlie all of life, and constrain what can and cannot occur during biological evolution, we can describe how life works without referring to them, in much the same way that we can describe what a painting looks like without referring to the absorption spectra of its pigments.

What *is* central to the origin of Earth life is the history of the universe—the cosmic dynamics that yielded our star, our planet, and the atoms that form living things. We can tell the story sparingly, without pausing to define terminology, allowing the flow of events to suggest the enormous times and distances involved.

THE UNIVERSE STORY

The observable universe is about fifteen billion years old. In the beginning, everything that is now that universe, including all of its space, was concentrated in a singularity, maybe the size of a pinhead, that was unimaginably hot (at least 100,000,000,000,000,000,000,000,000,000,000 degrees) and unimaginably dense. It all let loose during an event called the Big Bang, a misleading term in that there wasn't really an explosion. What happened was that the compacted space expanded very rapidly, carrying everything else along with it.

During the first three minutes of this expansion, all sorts of high-energy physics took place that yielded the current tally of subatomic particles in the universe, including protons, neutrons, and electrons. Some of the protons and neutrons fused to form helium ions, and random clumps developed in the expanding material so that it was not perfectly homogeneous. And then things started to settle down, with the space continuing to expand and cool until, after several hundred thousand years, temperatures were low enough that the protons and helium ions could acquire electron shells and become stable hydrogen and helium atoms. The expansion continued for another 15 billion years, yielding the present observable universe, 1,000,000,000,000,000,000,000,000 miles in diameter. Whether it will continue to expand or start to contract back again (the Big Crunch) is one of the many unknowns of cosmic evolution.

Because the early hydrogen and helium atoms were distributed inhomogeneously in the expanding space, close neighbors tended to move closer and then closer together, attracted by gravity. The result was that the universe became "lumpy," with vast gaseous clouds scattered here and there, occasionally colliding and merging with one another. These protogalaxies then differentiated, and continue to differentiate, into billions of galaxies, each giving rise to billions of stars.

A star starts out as a gaseous cloud, about three-quarters hydrogen and one-quarter helium. The atoms are brought together by gravitational attraction and, as they fall closer together, they speed up until the temperature is so high that they are stripped of their electrons and the hydrogen nuclei start to fuse, forming helium ions. These fusion reactions release heat, causing the gas to expand and counterbalancing its tendency to contract. As a result, the star stabilizes in temperature and size, often for billions of years, burning its hydrogen fuel.

Once the hydrogen begins to run out, the rate of nuclear fusion slows down and the gases no longer expand as readily. As a result, the star begins to contract again, eventually becoming so dense and hot that its helium nuclei start to fuse together, forming larger nuclei like carbon, oxygen, calcium, and other "light" elements of the periodic table.

What happens next depends on the size of the star. A small star becomes unstable at this stage and puffs away its outer layers, seeding the galaxy with its newly minted light elements and leaving behind a remnant known as a white dwarf. A giant star keeps collapsing, getting hotter and hotter and forming heavier and heavier nuclei until it starts to make iron, which it can't burn. When a critical amount of iron accumulates, the core of the star is crushed by gravity into what is called a neutron star, and the shock waves generated by the crushing process cause a huge explosion in the star's outer layers—a supernova. Very heavy nuclei, including radioactive elements like uranium, are created during the supernova phase, and all the new kinds of nuclei are released into gaseous clouds where they cool, acquire electrons, and become atoms.

The gaseous clouds now go on to aggregate into second-generation stars that are more complex than their predecessors because they include some of the new kinds of atoms. The second-generation stars proceed to burn their hydrogen and collapse, forming more new elements in the process, and the released detritus then reaggregates into third-generation stars that are yet again more complex. Such birth-and-death stellar cycles are apparently destined to continue for billions of years into the future.

THE EARTH STORY

So now we can look at our own context. The Milky Way is a medium-sized galaxy, and the Sun, located in one of its spiral arms, is a second- or third-generation medium-sized star that formed from the atoms released by a nearby supernova. The Sun has existed for about 4.5 billion years and has enough hydrogen to burn for another 5 billion years or so. During its terminal phases it is expected to become so hot that the Earth will turn into a cinder.

While the Sun was forming, some of the surrounding material assembled into small aggregates that grew and collided and merged with one another and eventually stabilized as its orbiting planets, moons, and comets. Importantly, some of these aggregates, including what is now Earth, contained generous quantities of the atoms spewed out by supernovae: These include the iron and radioactive elements that form the Earth's broiling core, the silicon that forms its crust, and the carbon, oxygen, nitrogen, and other elements that are essential for life. Moreover, comets colliding with the young Earth provisioned it with yet more atoms from distant supernovae, and also brought in a great deal of water in the form of ice. Gases trapped in the Earth's interior were released through fissures and volcanos and became trapped by gravity to form the early atmosphere, and the floating surface settled into large masses that drift and crash into one another in continuous geological activity, defining and redefining the continents and ocean basins. After about half a billion years of consolidation, the physical conditions on Earth became such that life could originate and continue.

Reflections

I've had a lot of trouble with the universe. It began soon after I was told about it in physics class. I was perhaps twenty, and I went on a camping trip, where I found myself in a sleeping bag looking up into the crisp Colorado night. Before I could look around for Orion and the Big Dipper, I was overwhelmed with terror. The panic became so acute that I had to roll over and bury my face in my pillow.

* All the stars that I see are part of but one galaxy.
* There are some 100 billion galaxies in the universe, with perhaps 100 billion stars in each one, occupying magnitudes of space that I cannot begin to imagine.
* Each star is dying, exploding, accreting, exploding again, splitting atoms and fusing nuclei under enormous temperatures and pressures.
* Our Sun too will die, frying the Earth to a crisp during its heat-death, spewing its bits and pieces out into the frigid nothingness of curved spacetime.

The night sky was ruined. I would never be able to look at it again. I wept into my pillow, the long slow tears of adolescent despair. And when I later encountered the famous quote from physicist Steven Weinberg—"The more the universe seems comprehensible, the more it seems pointless"—I wallowed in its poignant nihilism. A bleak emptiness overtook me whenever I thought about what was really going on out in the cosmos or deep in the atom. So I did my best not to think about such things.

But, since then, I have found a way to defeat the nihilism that lurks in the infinite and the infinitesimal. I have come to understand that I can deflect the apparent pointlessness of it all by realizing that I don't have to seek a point. In any of it. Instead, I can see it as the locus of Mystery.

* The Mystery of why there is anything at all, rather than nothing.
* The Mystery of where the laws of physics came from.
* The Mystery of why the universe seems so strange.

Mystery. Inherently pointless, inherently shrouded in its own absence of category. The clouds passing across the face of the deity in the stained-glass images of Heaven.

- The word God is often used to name this mystery. A concept known as Deism proposes that God created the universe, orchestrating the Big Bang so as to author its laws, and then stepped back and allowed things to pursue their own course. For me, Deism doesn't work because I find I can only think of a creator in human terms, and the concept of a human-like creator of muons and neutrinos has no meaning for me. But more profoundly, Deism spoils my covenant with Mystery. To assign attributes to Mystery is to disenchant it, to take away its luminance.

I think of the ancients ascribing thunder and lightning to godly feuds, and I smile. The need for explanation pulsates in us all. Early humans, bursting with questions about Nature but with limited understanding of its dynamics, explained things in terms of supernatural persons and person-animals who delivered the droughts and floods and plagues, took the dead, and punished or forgave the wicked. Explanations taking the form of unseen persons were our only option when persons were the only things we felt we understood. Now, with our understanding of Nature arguably better than our understanding of persons, Nature can take its place as a strange but wondrous given.

The realization that I needn't have answers to the Big Questions, needn't seek answers to the Big Questions, has served as an epiphany. I lie on my back under the stars and the unseen galaxies and I let their enormity wash over me. I assimilate the vastness of the distances, the impermanence, the *fact* of it all. I go all the way out and then I go all the way down, to the fact of photons without mass and gauge bosons that become massless at high temperatures. I take in the abstractions about forces and symmetries and they caress me, like Gregorian chants, the meaning of the words not mattering because the words are so haunting.

Mystery generates wonder, and wonder generates awe. The gasp can terrify or the gasp can emancipate. As I allow myself to experience cosmic and quantum Mystery, I join the saints and the visionaries in their experience of what they called the Divine, and I pulse with the spirit, if not the words, of my favorite hymn:

> Immortal, invisible, God only wise,
> In light inaccessible hid from our eyes,
> Most blessed, most glorious, the Ancient of
> Days,
> Almighty, victorious, thy great name we
> praise.
> Unresting, unhasting, and silent as light,
> Nor wanting, nor wasting, thou rulest in
> might,
> Thy justice like mountains high soaring
> above
> Thy clouds which are fountains of goodness
> and love.
> To all, life thou givest, to both great and

small;
In all life thou livest, the true life of all;
We blossom and flourish as leaves on the
tree,
And wither and perish, but naught changeth
thee.
Thou reignest in glory; thou dwellest in
light;
Thine angels adore thee, all veiling their
sight;
All laud we would render: O help us to see
'Tis only the splendor of light hideth thee.

      ❧ *Walter Chalmers Smith, 1867*

And then I wander back twenty-six centuries
to Lao Tzu and the first chapter of the Tao Te
Ching:

The Tao that can be told is not the eternal
Tao.
The name that can be named is not the eter-
nal name.
The nameless is the beginning of heaven and
earth.
The named is the mother of ten thousand
things.
Ever desireless, one can see the mystery.

Ever desiring, one sees the manifestations.
These two spring from the same source but
    differ in name; this appears as darkness.
Darkness within darkness.
The gate to all mystery.

11
Origins of Life

Every religion has an account of the origins of life. Most familiar to Western traditions is Genesis 1, a spare, poetic account of the six days of creation. The Pueblo Indians tell of a primordial home beneath a lake where humans, gods, and animals lived together while the earth above was still soft and "unripe." The Kagaba Indians describe a female Supreme Deity: "the mother of our songs, the mother of all our seed, bore us in the beginning of things.... She is the mother of the thunder, the mother of the streams, the mother of trees and of all things." Certain Hindu teachings speak of the Brahmanda, the cosmic egg from which all creatures came forth. The Yaruro of Venezuala tell of the water serpent Puana who created the world, his brother Itciai, the jaguar, who created water, and their sister, Kuma, wife of the Sun, who made the Yaruro people.

These are wonderful stories that still work for us as stories, but we recognize their cultural origins and their contradictions with our present understanding of what happened. When we look to the scientific account of Nature for an origins story, we find a very different kind of poetry. It goes something like this.

THE ORIGINS OF CHEMISTRY

In the beginning there was high-energy physics, but during the cooling of the universe we encounter the origins of chemistry. Chemistry allows atoms to form bonds with one another and hence associate into molecules; it also allows smaller molecules to associate into larger molecules. Like everything else, chemistry is reducible to physics, but chemistry can only take place under certain conditions. There are three conditions that are important to our story.

* Chemistry requires the flow of energy, from a source to a sink. The Earth has two important sources of energy: the Sun, of course, and also its own molten core, pulsing with radioactivity, that helps heat up the oceans and the continents. The energy sink is, ultimately, the universe itself, most of which is only a few

degrees above absolute zero. As energy flows, chemistry can occur.

* Chemistry cannot occur when atoms are so hot that they fall apart into subatomic particles. It also cannot occur when everything is so cold that the atoms are all locked up together as solids, like a rock. When temperatures are such that atoms and molecules can coexist in their various forms—solids, liquids, and gases—this is a sign that the system is enjoying energy flow and that chemistry can proceed.

* Some atoms are more likely to engage in chemistry than others. Helium, for example, exists only as helium, whereas carbon, hydrogen, nitrogen, oxygen, phosphorus, and sulfur— the Big Six in living systems—are poised to form bonds with one another under conditions of energy flow: hydrogens readily combine with oxygens to form molecules of water; carbons readily combine with oxygens to form carbon dioxide.

THE BUILDING BLOCKS

Two kinds of chemistry were needed to get life going: the chemistry that generates the so-called building blocks of life—water, carbon dioxide, and small molecules like formaldehyde, methane, and hydrogen sulfide—and the chemistry that allows these to associate into yet larger assemblies called biochemicals.

The molecular building blocks for Earth life are thought to have been generated in two factories. The first are tiny specks of matter, about the size of talcum particles, called interstellar dust because they form huge clouds of matter between the stars, soaking up the elements released from stellar catastrophes. Since they are floating out in space, the specks are inherently cold, but as they are pulsed with radiation from nearby stars, they heat and cool, heat and cool, and this energy flow allows hundreds of different kinds of complex molecules to form on their surfaces. As comets and meteors passed through interstellar clouds and then crashed into Earth, they brought with them large cargos of such building blocks to be used by incipient life.

The second building-block factories are thought to have been deep-sea hydrothermal vents. Water seeped into fissures in the Earth's mantle, heated up, and then circulated back into the cold oceans, creating an energy flow that again allowed for the synthesis of many kinds of molecules, perhaps on the surfaces of clay or iron particles.

So small but complex building blocks are thought to have accumulated in the waters of the new Earth from the time it formed, about 4.5 billion years ago, creating what is often called the "primal soup." For our purposes, the most important ingredients of the soup were three kinds of small molecules called sugars, amino acids, and nucleotides, with the nucleotides being of two sorts—ribonucleotides and deoxyribonucleotides. These are important because they prove to be the starting materials for all forms of Earth life.

And then, about 4 billion years ago, the second kind of chemistry got underway: the formation of biomolecules from these primal-soup building blocks.

THE BIOMOLECULES

Here our story is obscured by a large fig leaf. We don't yet know the sequence of events that gave rise to the first biomolecules and then to the first cell, and perhaps we never will. But we know a great deal about the end result, about the biomolecules and cells that have come to inherit the Earth, and this allows us to work backwards, to propose plausible, if perhaps not correct, scenarios for the generation of life as we know it, scenarios that serve to focus our attention on what it entails to be alive. Such a story might go something like this.

Imagine a puddle of primal soup on the new Earth containing a large number of ribonucleotide building blocks. There are four different kinds of ribonucleotides, called A, C, G, and U, and they have the tendency, left to themselves for a very long time under conditions of energy flow, to form chemical bonds with one another and produce chains of random length, much like those chains of cut-out paper dolls connected together by their hands and feet. These chains are called RNA. Left to themselves, ribonucleotides will string together in random sequences: one RNA molecule might read ACAUGCACUCA; the next might read GAGCCUAGCACUACG; and so on. No life yet.

Now, the moment. One of these molecules— let's say UAGCACGUAAACGUC—happens to have the ability to copy itself. RNA molecules with such properties are found in odd creatures that are alive today. It isn't important to our story to con- sider how they work. The important part is the result: At the conclusion of such a self-replication event, the puddle comes to contain two copies of UAGCACGUAAACGUC. The two can now self- replicate to yield four copies, and then eight, and then sixteen, such that if we were to return to the puddle after 1,000 years and sample its RNA population, we would find that most of the molecules are now UAGCACGUAAACGUC. The puddle would presumably still contain the other RNA molecules that had formed by chance, but since these are not self-replicating, they would now represent a vanishingly small minority of the total RNA present.

The early copying process would be anything but perfect—meaning that a given progeny molecule might, for example, carry a G rather than a U at the first position and read GAGCACGUA-AACGUC, a change known as a mutation. The mutation might cause the daughter molecule to lose its ability to self-replicate, in which case it would become a member of the small minority. But it might instead endow the molecule with the ability to self-replicate more rapidly, or more accurately, or both. Then, if we were to return to the puddle after another 1,000 years, we would expect to find that most of the molecules are now GAG-CACGUAAACGUC. The UAGCACGUAAAC-GUC versions would still be present, and replicating away as best they can, but they would no longer be the most prevalent. There would have occurred what is called natural selection for the faster and more accurate replicator.

But now, a crisis. During the course of making

all these RNA molecules, the puddle becomes depleted of its stock of ribonucleotide precursors—the A, C, G, and U building blocks that had provisioned the puddle. When the ribonucleotides run out, none of the RNA molecules, no matter how fast or accurate, can copy themselves. Everything stalls.

What happens next is that a mutant RNA molecule appears that is not only able to replicate itself, but is also able to carry the instructions for making more ribonucleotides, instructions encoded in units now known as genes. What this means, how it works, is the subject of later chapters. The important point here is that such an RNA molecule would have a huge advantage: It alone, of all the RNAs in the puddle, would be able to self-replicate because it alone could direct the synthesis of the ribonucleotides that are otherwise unavailable. It could, in effect, concoct its own primal soup, and natural selection would therefore favor its continuation.

Making more ribonucleotides and then allowing them to diffuse away into the puddle would not, however, be very efficient. A better strategy would be to surround the RNA in a membrane—a tiny bubble of lipid—such that when the ribonucleotides are synthesized, they stay inside the bubble, available for the next round of replication. Again postponing the question of how this works, we can say that further mutations generated an RNA molecule that carried the instructions—the genes—both for synthesizing ribonucleotides and enclosing them inside a membrane. At this point we are looking at the first cell: a membrane-enclosed self-replicating molecule capable of directing the synthesis of additional molecules, like membrane lipids and ribonucleotides, that make self-replication possible.

CELLS

There are good reasons to believe that cells with self-replicating RNA molecules were the first to inhabit the planet: The first world was apparently an "RNA world." But these cells have since vanished, or, rather, they have since evolved, into cells whose genes are encoded in DNA molecules, so we now live in a DNA world. DNA uses deoxyribonucleotides instead of ribonucleotides as precursors, and it is more stable than RNA, but the basic idea is otherwise the same: A long chain of deoxyribonucleotides carries genes whose molecular products make possible the replication of the chain.

Somewhere along the line, the genes encoded in RNA/DNA came to specify large molecules called proteins. Particularly important are proteins called enzymes, since they are responsible for getting the biochemistry inside the cell to proceed accurately and efficiently. How enzymes and other proteins work, and how DNA encodes their structure, will be considered in the next two chapters. Here we can stand back and take in the big picture.

The big picture is that a cell—and therefore life—must be able to construct itself, construct a cell, and then remember how to do it and pass the instructions on to daughter cells. We said that the initial building blocks in the primal soup were produced when pulses of energy from the sun or the Earth's interior made possible the chemistry of joining atoms together into molecules. We now say that this same kind of chemistry came to take place at moderate temperatures, and with increasing regularity and efficiency, inside tiny soap bubbles, with enzymes overseeing the biochemistry. The key role of DNA is to encode readable instructions for how to make the proteins and to pass these instructions along when it replicates.

Left out of this account is the critical question of how the chemistry is "driven"—that is, how energy is induced to flow through the cell at moderate temperatures and become trapped inside the biomolecules that are synthesized. Suffice it to say that along the way, cells first aquired the ability to extract energy from small molecules like hydrogen and hydrogen sulfide, and eventually developed the capacity to carry out photosynthesis—to capture energy from sunlight and transfer it into chemical bonds. Organisms that cannot do photosynthesis— like us—depend on the products of photosynthesis for survival: we ingest these products as food and then extract their energy in enzyme-mediated reactions collectively called metabolism.

PERSPECTIVE

For our origins story, then, two important points emerge. First, a system got thrown together, apparently quite by chance, that allows biomolecules to be sythesized by a sunlight-driven chemistry that is not at all left to chance. And, second, the instructions for constructing this system acquired the ability to be copied and inherited. That is, life emerged from nonlife. The stages that were traversed, the trials and errors, the near-extinctions, the struggles to recover, all these have been erased, supplanted by our intimate understanding of the ultimate winner, the first progenitor cell from whom all creatures flow.

Reflections

Life can be explained by its underlying chemistry, just as chemistry can be explained by its underlying physics. But the life that emerges from the underlying chemistry of biomolecules is something more than the collection of molecules. As we will see, once these molecules came to reside inside cells, they began to interact with one another to generate new processes, like motility and metabolism and perception, processes that are unique to living creatures, processes that have no counterpart at simpler levels. These new, life-specific functions are referred to as emergent functions.

The origin of life is but the first of many emergent functions we will encounter. A recent example is the emergence of self-awareness—our human ability to perceive the functioning of our own brains and call it "consciousness"—that in turn has given rise to the emergence of art and science and religious reflection. But we pause here to bear witness to the first discontinuity, the first biomolecules that were made again, and then again, and then again, from sets of RNA instructions. It was this that brought forth all the rest.

Emergence. Something more from nothing but. Life from nonlife, like wine from water, has long been considered a miracle wrought by gods or God. Now it is seen to be the near-inevitable consequence of our thermal and chemical circumstances.

But what about those circumstances? Does not some theology flow from the fact that the universe was so "right," and our planet was so "right," that life became inevitable? A line of reasoning called the Anthropic Principle states that since the laws of physics are perfect for the emergence of chemistry, and chemistry is perfect for the emergence of life, that it all must have been Designed so as to yield life in general and human life in particular. Had any of the laws of physics been anything other than what they are, the universe would have been very different, and perhaps not possible at all, and life as we know it would not have evolved.

True enough. But of course, all these things could just as well have happened by chance since, had they occurred any other way, we wouldn't be sitting here wondering about them. The inherent circularity of Anthropic-Principle arguments leaves me, in the end, theologically unsatisfied.

And so I once again revert to my covenant with Mystery, and respond to the emergence of Life not with a search for its Design or Purpose but instead with outrageous celebration that it occurred at all. I take the concept of miracle and use it not as a manifestation of divine intervention but as the astonishing property of emergence. Life *does* generate something-more-from-nothing-but, over and over again, and each emergence, even though fully explainable by chemistry, is nonetheless miraculous.

The celebration of supernatural miracles has been central to traditional religions throughout the millennia. The religious naturalist is provisioned with tales of natural emergence that are, to my mind, far more magical than traditional miracles. Emergence is inherent in everything that is alive, allowing our yearning for supernatural miracles to be subsumed by our joy in the countless miracles that surround us.

> I believe a leaf of grass is no less than the
>   journey-work of the stars,
> And the pismire is equally perfect, and a
>   grain of sand, and the egg of the wren,
> And the tree-toad is a chef-d'oeuvre for the
>   highest,
> And the running blackberry would adorn
>   the parlors of heaven,
> And the narrowest hinge in my hand puts
>   to scorn all machinery,

And the cow crunching with depress'd
  head surpasses any statue,
And a mouse is miracle enough to stagger
  sextillions of infidels.

&*Walt Whitman, Song of Myself, 1855*

𝒯𝒯𝒯

How Life Works

Even well into this century, even after it was clear that life works through myriads of chemical reactions and that the information needed to organize this chemistry is encoded in DNA molecules, there were scientists who continued to believe that there was "something else" about living systems, an élan vital, a vital force. Although scientists no longer invoke such a force—we are convinced that life emerges from biochemistry—the sense that there is "something else" about life is so widespread, so deeply rooted, that it almost seems instinctive.

The persistence of vitalism doubtless has many explanations, but I have come to believe that it persists as a bulwark to fend off reductionism. We are told that life is so many manifestations of chemistry and we shudder, a long existential shudder. And then we defend. We dig in our heels and say No! That can't be all it is! That does to life what astrophysics does to the night sky. Life reduced to its component molecules is life demeaned. Stop saying things like that!

For me, a helpful way to think about reductionism is to invoke what can be called the Mozart metaphor. A Mozart piano sonata is a wondrous thing, beautiful beyond belief, sonorous, resonant, transporting. But it is also about notes and piano keys. Mozart's magnificent brain composed the work, to be sure, and then he translated it into black specks on white paper to be translated into strings hit by tiny hammers. We can thrill to a piano sonata without giving a thought to its notes. But we can also open up a score and follow the notes, or play them ourselves, without having the music diminished or demeaned. It is another way of experiencing the whole and, indeed, the only way to have a full understanding of what the sonata entails and what Mozart had in his mind.

So let us go then, to life abstracted, life reduced to its most spare rendering, to the strings and hammers (the working parts) and the notes (the instructions). I can assure you that it is very beautiful where we are going, and not at all hard to understand. And after that we will return to the matter of alienation, and the response of religious naturalism.

PROTEINS

Living creatures are composed of cells. Most of the organisms on the planet are single cells, but some, like us, are made up of many different kinds of cells that cooperate to form a single organism. Each cell has a membrane around it, a thin film of lipid keeping the outside out and the inside in, and each cell contains the DNA instructions for its various activities.

If we could move inside a cell and start to watch the biochemistry (the working parts), it would become clear that it's all about shapes, particularly the shapes of proteins. Proteins are like jigsaw-puzzle pieces in three dimensions: They bristle with protuberances and pockets and long straight parts and tightly coiled parts, each part called a domain. The domains carry out the functions of the proteins, so it is important to understand how domains are put together.

When a protein is made, it starts out as a long chain of amino acids, one after the next, the same idea as the paper-doll chains of ribonucleotides in an RNA molecule except that the dolls are now amino acids. There are twenty different kinds of amino acids—twenty different shapes of paper dolls—each with its own properties: glycine is small and greasy; phenylalanine is bulky and greasy; aspartic acid is long and slender and carries a negative electrical charge. The instructions in a given gene dictate the sequence of amino acids in a given protein chain, so one kind of protein might start out with a glycine linked to a phenylalanine linked to an aspartic acid linked to another phenylalanine for a total length of 152 amino acids; a second kind of protein might start out with tryptophan linked to glycine linked to leucine for a total length of 433 amino acids.

Once one of these chains is synthesized, it folds up into its jigsaw-puzzle shape: Amino acids that prefer to be next to one another, like a group of greasy ones, might associate to form one domain; amino acids with negative charges might line up next to amino acids with positive charges to form a second domain; a bulky amino acid might cause a protuberant domain to stick out farther. This all happens spontaneously—the process is called self-assembly—and the result is a protein with a distinctive overall shape and size that displays a collection of very specific domains. A second chain with a different sequence of amino acids will self-assemble into a protein with a different overall size and shape and a different set of domains.

So you're doing a jigsaw puzzle, and you pick up a piece of sky that looks discouragingly like all the other pieces of sky, but you know to focus in on what's really important—the shape and size and placement of the protuberances that you know are going to have to fit exactly into the pockets of adjacent pieces. Proteins fit together in just this way to produce multiprotein complexes that perform many important functions in the cell, as we will see. Pockets are also crucial to the function of proteins called enzymes, since it is in these pockets that biochemistry occurs.

ENZYMES

When an enzyme folds into a jigsaw-puzzle shape, some of the pockets that self-assemble on its surface are not designed to form complexes with other proteins. Instead, they are shaped to interact with small molecules that the cell must manipulate chemically. Imagine an enzyme whose job is to take a molecule of the sugar glucose and a molecule of the sugar galactose and bring about the formation of a chemical bond between them to create glucose-galactose. This enzyme will have a pocket that is just exactly the right shape for a glucose molecule to fit into, and a second nearby pocket that's just the right shape for galactose, as diagrammed in Figure 1.

The next thing that happens is a bit hard to explain, but it's the critical step. Once both sugars are snuggled into their pockets, the enzyme is no longer the same. Instead of empty pockets it has full pockets, meaning that its amino acids no longer have the same neighbors that they used to have. In response, the enzyme changes its shape—it flips into a new configuration – and, in the process, the glucose and galactose come close enough together

∽ FIGURE 1 HOW AN ENZYME WORKS ∽ An enzyme (gray blob) with two "pockets," one filled with a glucose molecule and the other with a galactose molecule, changes its shape and brings the sugars close together so they can form a chemical bond (black bar that interconnects them).

to react with one another and form a chemical bond, as in Figure 1. The new glucose-galactose molecule then pops out, the enzyme resumes its original shape, and the process starts over again.

If you put many glucose and galactose molecules into a jar of water and wait a long time, two of them might spontaneously form a glucose-galactose bond. But if you add our enzyme, the jar will quickly fill up with glucose-galactose pairs. By offering its pockets so that the two sugars line up just so, and by then changing its shape so that the reactive parts of the sugars come together just so, the enzyme greatly enhances the probability that the chemical bond will form. The enzyme is said to catalyze the chemical reaction.

And that's basically all there is to biochemistry. Every cell is packed with thousands of different kinds of enzymes, each enzyme displaying a distinctive surface topology and each thereby able to catalyze one or several specific chemical reactions. Some enzymes catalyze the formation of chemical bonds, like our glucose-galactose example. Others catalyze the disruption of chemical bonds so that smaller molecules are generated from bigger ones. Yet another enzyme might take galactose-glucose and add a third sugar on to it, and then a fourth, until a long chain, a polysaccharide, is generated.

This same kind of operation generates long chains of amino acids (proteins), long chains of nucleotides (DNA and RNA), and long chains of glycerides (lipids), all of which are key cellular components.

BIOPHYSICS

Biochemistry is supplemented by biophysics. Particularly important to biophysics are protein assemblies called channels and pumps that span the cell membrane and determine which electrically charged ions (e.g., calcium, potassium, chloride) can cross the membrane and at what rate. Since some of these ions carry a positive charge and some a negative charge, their net distribution generates ion gradients: The inside of the cell is rendered more negative than the outside, for example, and contains more potassium and less sodium and calcium than the outside. These so-called electrochemical gradients are essential for a cell to function properly, and a cell quickly dies if they are disrupted.

CASCADES

Knowing this much, we can now step back and watch what's going on inside a cell as if we were watching a movie. As we watch, we realize that life proceeds as a series of shape changes. We might observe three proteins fitting together to produce a complex that has the correct shape to associate with some lipids in the cell membrane, forming a sodium channel. When the channel opens by changing its shape, the resultant influx of sodium causes an internal enzyme to change its shape such that certain pockets, previously buried in its interior, become exposed. These pockets are now available to catalyze some biochemistry, the products of which go on to induce another protein to change its shape, and so on. Such a sequence is called a cascade—literally a series of small waterfalls tumbling down a hill, top to bottom, start to finish.

Cascades also describe how a cell perceives things. Every cell membrane carries proteins called receptors. One domain of a receptor faces outside, into the environment; a second domain bridges the membrane; a third faces the cell interior. The outer face carries a pocket that is correctly shaped to fit some molecule in the external world—for example, the membranes of the cells in your nose are studded with receptors shaped to fit particular odorants. Figure 2 depicts such an olfactory receptor, whose rectangle-shaped outer pocket is shown associating with a rectangular odorant molecule. Just as we saw with enzymes, this pocket-occupancy causes the receptor to change its shape, a shape-change that propagates through the membrane-spanning domain and creates a round-shaped pocket in the interior domain. A round intracellular molecule can snuggle into this new site and the receptor, now acting as an enzyme, catalyzes a change in the shape of the intracellular molecule (it acquires "ears") such that it can interact with yet another molecule and bring about a change in its shape. And so on, down the line, one shape change catalyzing the next until the organism experiences the odor. The signal—the presence of the odor—is said to set off a signal-transduction cascade: the receptor transduces the external signal into appropriate biochemistry/biophysics.

Even those of us experienced in thinking about biological cascades are astonished at their speed. All the events diagrammed in Figure 2, plus a great deal of brain-centered biophysics necessary to process and evaluate the odor, take place well within a second. We know this by experience, of course—we know how long it takes to smell something—but our intuition gives us no hint as to what is going on during this second.

If we back off and watch the action even more globally, we realize that the inside of the cell is set up to optimize the flowing of cascades. Proteins

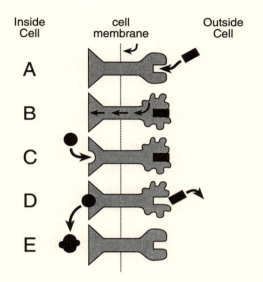

Inside Cell	cell membrane	Outside Cell

A

B

C

D

E

∞ FIGURE 2 HOW A MEMBRANE RECEPTOR WORKS ∞ A receptor protein, shaped like a wrench, is shown spanning a cell membrane (A). Its outer face contains a rectangle-shaped pocket designed to be occupied by a rectangular molecule from the outside world. When this occurs (B), the receptor is induced to change its shape (squiggly lines), and this change translates along the length of the protein (arrows) to create (C) a new round pocket exposed to the inside of the cell. A round intracellular molecule can fit into this pocket (D) and the receptor, now acting as an enzyme, catalyzes a shape change in the round molecule—it acquires "ears" (E)—such that it has the correct shape to interact with the next intracellular molecule in the signal-transduc-

destined to interact with one another are endowed with domains, called addresses, that target the proteins to the same cellular location. Each destination proves to be optimal for particular biochemical reactions—some locales are fatty and stolid, some are aqueous and dynamic, some are acidic, some are loaded with calcium—and each is delimited by its own boundary, often an intracellular membrane. Some of these compartments stand alone like fortresses, but many undergo elaborate branching and anastomosis, mixing their products and then separating again. Thus, a cell is like a community, its inner workings segregated into interacting compartments, its outer membrane defining its interactions with the rest of the world.

THE INSTRUCTIONS

Every organism is endowed with a complete set of instructions for how to make all its proteins, and these instructions are copied so that more organisms can be produced. The instructions, the memory, are stored in DNA, which uses a code to specify the sequences of amino acids —the strings of paper dolls—that then self-assemble as three-dimensional enzymes or receptors or channels. Each sector of DNA that encodes a protein is called a gene, and the collection of all the genes necessary to specify an organism is called its genome (the human genome, for example, contains about 100,000 genes). For a lineage to continue, the entire genome must be replicated and then transmitted to the next generation, much as we now have numerous copies of the complete works of Mozart. We will have much more to say about genes and genomes in the next chapter.

THE SONATA

The notes are hammered by the piano keys and out flows the sonata. Our attention moves from the DNA sequence to the rose blossom and back down again to the intracellular membrane compartments swollen with red pigment to give the petals their flush. We reduce, and then we synthesize, and then we find another occasion to reduce. How did Mozart generate that modulation into B-flat? Ah, with that chord. How lovely.

Reflections

Reductionism presents a hierarchy of truths. The hierarchy is not about parts *versus* wholes; it's about sub-wholes: when you look down, it feels like you're looking at the whole thing, but when you look up, you realize you've only been looking at a part. To be sure, the wholes display properties and behaviors—emergent functions—that the parts cannot, but this does not mean that the parts are somehow irrelevant, or somehow untrue.

We can recite these words, and the Mozart metaphor, and we can develop a deep understanding of, and admiration for, the notes and the strings and the keys of life. As a cell biologist immersed in this understanding, I experience the same kind of awe and reverence when I contemplate the structure of an enzyme or the flowing of a signal-transduction cascade as when I watch the moon rise or stand in front of a Mayan temple. Same rush, same rapture.

But all of us, and scientists are no exception, are vulnerable to the existential shudder that leaves us wishing that the foundations of life were something other than just so much biochemistry and biophysics. The shudder, for me at least, is different from the encounters with nihilism that have beset my contemplation of the universe. There I can steep myself in cosmic Mystery. But the workings of life are not mysterious at all. They are obvious, explainable, and thermodynamically inevitable. And relentlessly mechanical. And bluntly deterministic. My body is some 10 trillion cells. Period. My thoughts are a lot of electricity flowing along a lot of membrane. My emotions are the result of neuro-transmitters squirting on my brain cells. I look in the mirror and see the mortality and I find myself fearful, yearning for less knowledge, yearning to believe that I have a soul that will go to heaven and soar with the angels.

William James: "At bottom, the whole concern of religion is with the manner of our acceptance of the universe."

The manner of our acceptance. It can be disappointed and resentful; it can be passive and acquiescent; or it can be the active response we call assent. When my awe at how life works gives way to self-pity because it doesn't work the way I would like, I call on assent—the age-old religious response to self-pity, as in "Why, Lord? Why This? Why ME?" and then, "Thy Will Be Done."

As a religious naturalist I say "What Is, Is" with the same bowing of the head, the same bending of the knee. Which then allows me to say "Blessed Be to What Is" with thanksgiving. To give assent is to understand, incorporate, and then let go. With the letting go comes that deep sigh we call relief, and relief allows the joy-of-being-alive-at-all to come tumbling forth again.

Assent is a dignified word. Once it is freely given, one can move fluidly within it.

IV

How an Organism Works

The Mozart metaphor began with our delight in the sonata. We then acknowledged that the music was the consequence of notes and hammers. Invoking the concept of emergence, we can say that the music emerges from the notes. But there's an intermediate level of emergence as well. From the notes emerge chords and phrases and tempos and melodies, and from these emerge the sonata as a whole. In the middle there are musical patterns.

In biology it is the same. The biochemistry and biophysics are the notes required for life; they conspire, collectively, to generate the real unit of life, the organism. The intermediate level, the chords and tempos, has to do with how the biochemistry and biophysics are organized, arranged, played out in space and time to produce a creature who grows and divides and is. In the middle there are biological patterns.

In order to give assent to who we are, we need to understand who we are. We need to understand how biochemistry and biophysics generate the patterns that make us these pulsating organisms. To do this we can start with an amoeba, for once there is understanding of how an amoeba is put together, most of the story falls into place. We will next consider the ways that our patterning is different from an amoeba. And finally we will ask how we can locate the sacredness of our individual selves within this biological context.

GENE EXPRESSION

The genome of an amoeba contains all the genes, all the instructions, for making all the proteins needed to generate that particular kind of an amoeba. Each protein has an address that tells it where to go in the cell, and each protein has a shape that dictates with what it will associate and the biochemical/biophysical consequences of that association. Fine. But the trick to crafting an amoeba's life out of this arrangement is to do one more thing: regulate the pattern of gene expression.

To understand what this means, we first need to understand how a gene is expressed, and then we can look at how gene expression is regulated.

As noted earlier, the original genes in the

primeval puddle were apparently encoded in RNA molecules, whose ribonucleotide building blocks are abbreviated A, C, G, and U, whereas most modern genes are encoded in DNA, whose building blocks, the deoxyribonucleotides, are abbreviated A, C, G, and T. Each of the nucleotides has a distinctive shape, and they line up in paper-doll chains (e.g., ATGCACTGGCCC...).

A gene is a segment of a such a DNA molecule that contains the instructions for producing a protein. Specifically, the sequence of nucleotides in a DNA chain dictates the sequence of amino acids in a protein chain. The basic idea is that each amino acid is coded for by a set of three nucleotides called a codon. The codon specifying the amino acid methionine is ATG, the codon specifying histidine is CAC, the codon specifying tryptophan is TGG, and the codon specifying proline is CCC. So, if methionine, histidine, tryptophan, and proline are the first four amino acids in the amino-acid chain composing the enzyme lactase, then the lactase gene will begin with the sequence ATG,CAC,TGG, CCC... and continue along to the other end.

The codon sequences in a gene are converted into the amino-acid sequences of a protein by two complex biochemical processes called transcription and translation, during which several elegant pieces of cellular machinery recognize the shapes of the codons and catalyze the formation of chemical bonds between the corresponding amino acids. When a DNA sequence is converted into an amino-acid sequence, we say that the gene has been expressed, that its instructions for making a protein have been read.

REGULATION OF GENE EXPRESSION IN TIME

Now we can look at the regulation of gene expression. A gene is an instruction for making a protein, and a cell has the option to express that gene and hence contain that protein, or not express that gene and hence lack the protein. It also has the option to express the gene often, and hence have a lot of the protein, or express it rarely and hence have little. These decisions involve domains of DNA called promoters.

Our lactase gene starts out ATG,CAC,-TGG,CCC... and continues to the end of the protein-coding sequence. If we go to the initial ATG and move leftward along the DNA, we encounter a string of nucleotides constituting the promoter, a portion of which might read ...CATTAAGCATAG, followed by our ATG,CAC, TGG, CCC.... The promoter does not encode amino-acid sequences. Instead, it constitutes the control module that determines whether the lactase gene is expressed or not and, if so, at what rate. Once again, this is all done by shape. The CATTAAGCATAG promoter sequence has a distinctive shape, and proteins will bind to it if they have the correct pockets and protuberances to do so. If a so-called activator protein recognizes the promoter's shape and binds to it, then the expression of the gene is activated and lactase is made. If a repressor protein binds to the promoter, then expression is blocked and lactase is not made.

Let's look at an example. The lactase enzyme works by breaking down (metabolizing) the sugar lactose. The lactase gene of an amoeba is ordinarily "switched off" because amoebas don't usually encounter much lactose in their environment. If, however, an amoeba crawls into a lactose-rich puddle, then its lactase gene is "switched on," lactase is made, and the lactose can be used as a food source. How does this work?

Embedded in the amoeba's cell membrane are lactose receptor proteins that bind to molecules of lactose in the puddle. Binding causes the receptors to change their shape, much as we saw earlier for olfactory receptors (Figure 2), and this sets off a signal-transduction cascade that eventually brings about a shape change in an activator protein. The activator is now able to recognize, and bind to, the lactase promoter, and the lactase gene is expressed. When an adequate number of lactase enzymes have been produced to metabolize the lactose being ingested, a second signal-transduction cascade is triggered that leads to a reversal of the activator's shape change so that it can no longer bind to the promoter and activate the gene. The gene is "switched off." A repressor protein may now come along and bind to the promoter, thereby inhibiting the gene's expression more definitively.

An amoeba produces hundreds of different kinds of gene activators and gene repressors. Each is a protein, meaning that each is encoded by its own gene that possesses its own promoter subject to its own set of activators and repressors. So where does all of this end?

It all gets coordinated by a vast series of feed-back loops that are often very complicated. For example, the activator protein capable of switching on the lactase gene may also be able to combine with another protein to switch off a second gene. When the second gene stops being expressed, a signal transduction cascade is stimulated and a third gene is activated whose product, a repressor, switches off the gene for the lactase activator. The diagrams for all this look like wiring diagrams for a vastly elaborate machine which is, of course, what an amoeba is. And any failure in the system can have important consequences. If the lactase promoter sequence were to undergo a mutation such that just one nucleotide were changed (CATT<u>G</u>AGCATAG), the activator might be unable to recognize its shape and the amoeba would be unable to feed on lactose.

INTERNAL CLOCKS: THE CELL CYCLE

The amoeba can switch genes on and off in response to changes in the environment, as in our lactose/lactase example. In addition, important sets of genes are regulated internally, marching to their own drummer, a good example being the genes that govern what is called the cell cycle.

An amoeba feeds and grows in size until a decision is made to copy the entire genome (DNA replication) by an elaborate enzymatic process. Once replication is finished, a second decision is made that allows the amoeba to divide in two (mitosis), with one genome copy going to one daughter organism and the other to the other. And then the cell cycle starts over again, each daughter first growing, then replicating its DNA, then dividing by mitosis. Each "grow," "replicate," and "divide" decision is bracketed by a large number of subdecisions, and all are dictated by changing patterns of gene expression.

Let's say we start watching the cycle when the cohort of "replicate" genes has just switched on. Some of their protein products mediate DNA replication directly; others repress genes that were active during the growth stage; others activate "divide" genes that must next be expressed to kick off mitosis. Some of the resultant "divide" proteins participate in mitosis directly while others serve to switch off the "replicate" genes. And then, once cell division is completed, the "divide" genes are repressed and "grow" genes are activated to complete the cycle.

The time it takes for a cell cycle to elapse can be influenced by the environment—a well-fed amoeba cycles faster than a poorly fed one—but cell cycles have an inherent life of their own. Indeed, once the first cycle was traversed, the engine has never stopped: Cell cycles have been continuously running, continuously generating daughter organisms, for billions of years.

REGULATION OF GENE EXPRESSION IN SPACE
We need but one more variation on this theme to take us from amoebas to humans. Perhaps a billion years ago, a set of genetic instructions changed such

that the two daughter cells didn't crawl away from each other after mitosis, but instead stayed together to form a two-celled organism. And then the instructions changed again such that there were also four-celled organisms, and so on.

The key invention to get this to happen was sex, the subject of a later chapter, but a critical innovation also occurred in the patterning of gene expression. In addition to regulating the expression of genes in *time*—switching on a lactase gene after lactose is perceived, or switching off "replicate" genes after the genome has been copied—organisms were now also able to regulate gene expression in *space*. Our two-celled organism, for example, might be programmed to switch on a set of light-detection genes in cell #1 and a set of motility genes in cell #2. We would now have an organism in which the motile cell is found pushing its light-sensing cell ahead of it like a tiny eye. The four-celled organism might expand on this idea, having one light-detecting cell specialized for blue light and a second specialized for yellow light.

Multicellular. This is what I am, we are. In humans, more than a trillion daughter cells remain together to form an organism. Each cell possesses the full set of genetic instructions for making a human being, but only some of these instructions are read: blood cells switch on the genes encoding hemoglobin but never express the genes encoding the hair protein keratin; hair-follicle cells, in contrast, switch on keratin genes but never hemoglobin genes. Moreover, each type of cell is cycling at its cell-specific rate—certain kinds of cells in my gut divide twice a day, whereas my liver cells divide only once a year and my nerve cells don't divide at all—and these patterns generate an organism with a stable size and shape.

It all gets set up during embryogenesis, when a fertilized egg divides in 2 and then 4 and then 8 and eventually reaches 10 trillion, with each set of daughter cells making decisions that influence decisions that influence decisions, each decision marked by a set of genes switching on or off in particular cells at particular times in the developmental sequence. And then, once it's set up, once the correct cells form correct tissues and organs, the whole thing goes. Brains coordinate muscles, hormones coordinate metabolism and fear and conception, hearts pump blood and kidneys filter it.

So now we can return to Mozart. Patterns of gene expression are to organisms as melodies and harmonies are to sonatas. It's all about which sets of proteins appear in a cell at the same time (the chords) and which sets come before or after other sets (the themes) and at what rate they appear (the tempos) and how they modulate one another (the developments and transitions). When these patterns go awry we may see malignancy. When they change by mutation we can get new kinds of organisms. When they work, we get a creature.

Reflections

At the baptismal ceremony at my church, the parents hand the shining baby over to the minister. He looks down lovingly, dips his hand in the water, touches the luminous little head three times, and says, "I baptize you in the name of the Father, and the Son, and the Holy Spirit. You are a child of the Covenant, called by name, cherished, known, blessed by the grace of God."

Called by name. This brand new creature, called by name. I gasp every time I hear the words. The self, the soul: created, known, immortalized, saved. I was taught to sing "Jesus loves *me* this I know," and to "pray the Lord *my* soul to take." What do I do with my yearning to be special in some ultimate sense?

I have come to understand that the self, my self, is inherently sacred. By virtue of its own improbability, its own miracle, its own emergence.

I start with my egg cell, one of 400,000 in my mother's ovaries. It meets with one of the hundreds of millions of sperm cells produced each day by my father. Astonishing that I happen at all, truly astonishing. And then I cleave, I gastrulate, I implant, I grow tiny fetal kidneys and a tiny heart. The genes of my father and the genes of my mother switch on and off and on again in all sorts of combinations, all sorts of chords and tempos, to create something both eminently human and eminently new. Once I am born, my unfinished brain slowly completes its maturation in the context of my unfolding experience, and during my quest to understand what it is to be a person, I come to understand that there can be but one me.

And so I lift up my head, and I bear my own witness, with affection and tenderness and respect. And in so doing, I sanctify myself with my own grace. To the extent that I know myself, I am known. My yearning to be Known is relegated to the corridors of arrogance, and I sing my own song, with deep gratitude for my existence.

With this comes the understanding that I am in charge of my own emergence. It is not something that I must wait for, but something to seek, something to participate in achieving, something to delight in achieving. As my self-knowledge deepens and evolves, I find myself in spiritual alignment with the voice that calls to us so poignantly from the Book of Hours (Sarum, 1514):

> God be in my head and in my understanding,
> God be in mine eyes and in my looking,
> God be in my mouth and in my speaking,
> God be in my heart and in my thinking,
> God be at mine end and at my departing.

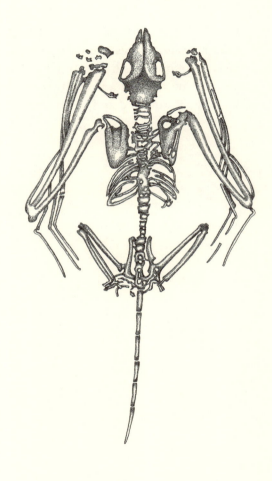

𝒱

How Evolution Works

Life, we can now say, is getting something to happen against the odds and remembering how to do it. The something that happens is biochemistry and biophysics, the odds are beat by intricate concatenations of shape fits and shape changes, and the memory is encoded in genes and their promoters. We read the notes, we hear the emergent chords and harmonies, and we marvel at the emergent musical experience.

To understand how this all came to be, we can compare the history of life with the history of music. The music of a composer like Brahms did not spring from his brain *de novo*. A good musicologist can go through a Brahms score and point out a Bach-like fugal texture here, a Handelian cadence there, a Hungarian folk melody somewhere else. As Brahms composed, bits of the old were woven together with the new to generate the next musical legacy.

In the history of life, the evolution of life, it is the same. A good biochemical idea—a protein domain that binds well to a promoter, a channel that's just the right size for a calcium ion— gets carried along through time, tweaked and modulated to best serve the needs of the current composition/organism but recognizable throughout evolutionary history. These conserved ideas combine with novelty to generate new directions, new ways of negotiating new environmental circumstances.

Evolution can be minimally defined as changes in the frequencies of different sets of instructions for making organisms. So, to understand evolution, we need to understand how the instructions become different (mutation) and then how the frequencies of those instructions are changed (natural selection). And then we can take in how this process has created a deeply interconnected web of life.

MUTATION

Genetic instructions are changed by mutation. A mutation is simply a change in the sequence of nucleotides in a genome, a concept we first encountered when we considered the first RNA molecules in the puddles of the early Earth. A mutation may arise as an uncorrected error during DNA replication, or it may arise as the consequence of chemical or physical damage to the genome from environmental agents. If we go back to our earlier example of the lactase gene that starts out ATG, CAC, TGG, CCC..., then a mutant form might read ATG, CAC, TGT, CCC and the protein would have cysteine (TGT) rather than tryptophan (TGG) as its third amino acid.

This mutation can have one of three effects. If the tryptophan is critical for lactase to work and cysteine is the wrong shape for that purpose, then the mutation is said to be deleterious. If the cysteine allows the enzyme to work better, then the mutation is beneficial. And if the original and the new enzyme work pretty much the same—perhaps position #3 doesn't participate in forming an important domain—then the mutation is said to be neutral. Mutations in promoters will also have deleterious, beneficial, or neutral consequences depending on which nucleotide(s) are altered: An activator or repressor may recognize a mutated promoter sequence less well, better, or about as well as its predecessor.

Important for the generation of novelty are mutations, known as duplications, that occur when the DNA replication enzymes get confused and copy the same gene twice. Were this to happen with our lactase gene, then one of the duplicates can continue to cover the lactase-producing function, meaning that the other duplicate is in effect a free agent: As it accumulates mutations over time, it generates new chains of amino acids that fold into new shapes. Should one of these new proteins serve the organism in some new way, the lineage can be said to have acquired a novel gene.

Mutations are random: Every nucleotide in the genome is equivalently vulnerable to change; every gene sequence is equally vulnerable to duplication. By contrast, the fate of mutations is not at all random. Instead, each new gene, promoter, or duplication is subject to very discriminating, very purposeful acts of selection.

NATURAL SELECTION

Natural selection asks two questions: Does the new protein or promoter work better, worse, or the same as the old one? And, how important is this difference to the organism? So, for an amoeba dependent on lactose as its food source, a deleterious lactase mutation will likely be lethal and the new gene will fail to spread, whereas a beneficial mutation may allow it to grow and divide more rapidly and hence the new gene may come to be more prevalent than the old one. For amoebae able to use an alternate food source like maltose, on the other hand, mutant lactase genes will "drift" along through many generations without being a focus of selection. If, however, the environment changes so that maltose stops being available and lactose is the only food option, then the selection pressures change completely. Now the quality of the lactase genes becomes critical, and organisms with the most effective lactase activity will prevail.

So, mutations change the quality of genes. Natural selection changes the frequency of these genes. And the ongoing, underlying fact is that the process is totally dependent on context. Evolution is contingent on the environmental circumstances in which it is occurring: lactose or maltose, warm or cold, prey or predators, wet or dry. These are the agents that call the shots.

COMPLEX TRAITS AND CUMULATIVE CHANGE

The dual dynamics of evolution—mutation and natural selection—are most readily understood by thinking about a single gene like the lactase-encoding gene. However, the traits that define an organism—its motility, its perception of odors, its metabolism, its embryology—are dictated by sets of genes whose products interact in space and in time. Such complex traits are therefore the true substrates of evolution.

To illustrate how a complex trait might get started and then evolve, we can reconstruct the kinds of events that might have been involved in giving rise to the modern flagellum of the modern bacterial cell. A flagellum sticks out from the bacterial surface and rotates around in the water like a tiny propeller, pushing the bacterium forward or backward. Each flagellum consists of a long fibrous appendage anchored into a flywheel motor that spins round and round in the bacterial membrane. The whole assemblege is built of some twenty different kinds of proteins and hence is dictated by twenty different genes.

We can begin with a primitive bacterium that lacks a flagellum. One of its genes duplicates, and mutations accumulate randomly in one of the duplicates until its mutant protein product adopts a shape that happens to allow it to bind to a membrane channel. This channel transports hydrogen ions and therefore controls the acidity of the cell. When the new protein binds to the channel, the channel acquires two new properties: it can now rotate in the plane of the membrane, and it can transport acid more efficiently (whether this is because it rotates or for some other reason is quite beside the point). Natural selection proves to favor bacteria with more efficient acid transport, and the new "rotator gene" therefore spreads through the lineage. There will also be positive selection for any additional mutations in the channel gene or in the rotator gene that improve the rotator/channel interaction and hence improve the acid transport process.

A key feature in the generation of novelty lies in the parentheses of the last paragraph. What is under selection at the outset is the ability to transport acid. The rotation of the channel is quite beside the point—it is an "unintended consequence" of the improvement in transport kinetics. But now, as a heritable trait, the rotation becomes a putative substrate for natural selection.

Let's say that a second gene now happens to duplicate, a gene that dictates the structure of some fibrous protein. Mutations accumulate in one of the duplicates until it comes to specify a fibrous protein with the correct shape to bind to the rotator-channel complex and stick out into the water—the first flagellum. Since the flagellum allows the bacterium to move and thus obtain food more readily, the new flagellum gene spreads in the lineage and comes to prevail. Additional mutations will accumulate in the flagellum gene that optimize the interaction of the flagellum with the rotator-channel complex. Moreover, since the ability to rotate is now under selection as well as the ability to transport acid, new mutations in the rotator gene will be favored if they improve the speed and efficiency of rotation. Also favored will be mutant versions of other genes whose products bind to the complex and improve its propeller properties. The outcome, after hundreds of millions of years of evolution, is the modern bacterial flagellum with twenty proteins dedicated to its optimal operation. Were we able to examine bacteria fifty million years from now, we might find that their flagellar motors are built from thirty proteins, the ten new ones confering additional speed and flexibility, or we might find motors built from only fifteen proteins, each so optimized for flagellar function that five of the "originals" became obsolete.

The general principle here, then, is that evolution produces *cumulative* change. New structures, new protein molecules, do not leap into existence fully formed. Rather, they appear as slightly modified versions of previously existing proteins that were less efficient, or served a different function, or served the same function under different conditions. Different parts of old systems get connected up in new ways; old proteins are modified to make new ones. Increasing complexity entails selections of selections of selections.

NOVELTY VERSUS CONSERVATION

When we think about the evolution of life, most of us marvel at all the new ideas, the new kinds of feathers and flowers and mating rituals. But at the gene level, evolution proves to be remarkably conservative. As we noted earlier, once a gene sequence has arisen that encodes a particularly useful protein domain—a domain that binds well to DNA, or a domain that can catalyze the transfer of phosphate groups onto proteins, or a domain that allows cells to find one another in an embryo—the sequence shows up again and again, in different guises in different genes in different lineages. The French word for this is bricolage: the construction of things using what is at hand, the patchwork quilt.

As a consequence of bricolage, a great deal of homology exists between the genes of all modern-day organisms, reflecting the fact that we have all evolved from the same common ancestor and have moved through evolution manipulating the same basic sets of protein domains. Gene families—families of ion channels, families of gene activators, families of DNA replicases—are found everywhere, deep in the sea and underneath rocks and flying about in the sky.

Similarly, once a complex trait is established, like a signal transduction cascade or a metabolic pathway or an embryonic induction, it also tends to be used again and again, with appropriate embellishments to suit the circumstances. Thus, many of the genes governing the yeast cell cycle have homologues in the human genome; indeed, a yeast cell carrying a deleterious mutation in one of these genes, and hence unable to grow, can be "rescued" if it is provided with the homologous human gene.

So, all the creatures on the planet today share a huge number of genetic ideas. Most of my genes are like most gorilla genes, but they're also like many of the genes in a mushroom. I have more genes than a mushroom, to be sure, and some critical genes are certainly different, but the important piece to take in here is our deep interrelatedness, our deep genetic homology, with the rest of the living world.

Reflections

Fellowship and community are central to the religious impulse. Children of Israel. United in Christ. Umma in Islam. A friend who was raised Roman

Catholic and who travels frequently to foreign cities tells me that she often seeks out the local church when she arrives, finding there the shared ritual, the known liturgy and prayers, the haven. Those of us who find a religious home feel deep affinity with those who have moved through with us and before us, congregating, including, supporting. We offer and receive sympathy and affection. The musicians sing their hushed responses or chant their solemn rhythms and we breathe together, sense our connectedness, heal.

Religion. From the Latin *religio,* to bind together again. The same linguistic root as ligament. We have throughout the ages sought connection with higher powers in the sky or beneath the earth, or with ancestors living in some other realm. We have also sought, and found, religious fellowship with one another. And now we realize that we are connected to all creatures. Not just in food chains or ecological equilibria. We share a common ancestor. We share genes for receptors and cell cycles and signal-transduction cascades. We share evolutionary constraints and possiblilties. We are connected all the way down.

I walk through the Missouri woods and the organisms are everywhere, seen and unseen, flying about or pushing through the soil or rummaging under the leaves, adapting and reproducing. I open my senses to them and we connect. I no longer need to anthropomorphize them, to value them because they are beautiful or amusing or important for my survival. I see them as they are; I understand how they work. I think about their genes switching on and off, their cells dividing and differentiating in pace with my own, homologous to my own. I take in the sycamore by the river and I think about its story, the ancient algae and mosses and ferns that came before, the tiny first progenitor that gave rise to it and to me. I try to guess why it looks the way it does—why the leaves are so serrated and the bark so white—and imagine all sorts of answers, all manner of selections and unintended consequences that have yielded this tree to existence and hence to my experience.

You do not have to be good.
You do not have to walk on your knees
for a hundred miles through the desert, repenting.
You only have to let the soft animal of your body
love what it loves.
Tell me about despair, yours, and I will tell you mine.
Meanwhile the world goes on.
Meanwhile the sun and the clear pebbles of the rain
are moving across the landscapes,

over the prairies and the deep trees,
the mountains and the rivers.
Meanwhile the wild geese, high in the clean blue air,
are heading home again.
Whoever you are, no matter how lonely,
the world offers itself to your imagination,
calls to you like the wild geese, harsh and exciting—
over and over announcing your place
in the family of things.

❧ Mary Oliver, 1986

Blessed be the tie that binds. It anchors us. We are embedded in the great evolutionary story of planet Earth, the spare, elegant process of mutation and selection and bricolage. And this means that we are anything but alone.

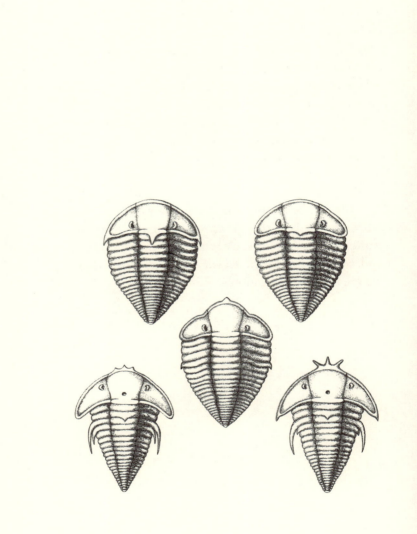

VI

The Evolution of Biodiversity

THE GENERATION OF BIODIVERSITY

Our description of how evolution works could apply to any planet in any solar system, any galaxy. Once there's a self-sustaining biochemical system that can replicate its own instructions, then any changes in the instructions that allow more copies of the system to be produced will become more prevalent. If ours were a perfectly uniform planet with a totally homogeneous environment, it would presumably come to be inhabited by a single kind of organism, maximally adapted to such a habitat, beavering away at its self-replication.

Happily for us, our planet is anything but homogeneous. Instead, it offers, and continues to generate, a seemingly endless diversity of environmental parameters: arid and humid, fresh and salty, aerobic and anaerobic, with and without other kinds of organisms. A collection of such parameters that generates the opportunity for habitation is called a niche. Organisms that attempt to populate a niche must be able to operate in that context; genes that improve this possibility will be selected for, and genes that hinder this possibility will be selected against. Thus there is no such thing as the "fittest" kind of organism. We can only talk about how an organism propagates in a given niche, how its life strategies have become adapted to that niche. It is no more or less fit than another kind of organism that has adapted to some other niche.

So because we have an endless array of niches, with tectonics and glaciations to stir things up in the long term, and tides and seasons and weather to modulate things in the short term, we have had an endless array of organisms. And what a windfall it has been! Minute and enormous, beautiful and hideous, enduring and evanescent, independent and parasitic. They occupy the most impossible (to our eyes) niches: ocean vents, arctic snows, desert cliffs, human eyelashes. They form long complex food chains and, in the process, provision our atmosphere with gases and our earth with soil and our biosphere with fixed carbon and nitrogen.

We know the story of the dinosaurs, of how they came and went. We will consider the evolution of ourselves in later chapters. Here we can walk through the overall history of life, thinking about what happened to all the genes and the creatures who carried them. And then we can think about our place in it all.

THE EVOLUTIONARY STORY

The common ancestor of life on Earth was a single-celled organism that sits in the center of Figure 3A (dotted circle), the three major branches of life extending out from it. This ancestor is unlikely to have been the first cell on Earth; doubtless there were many versions of cells that were struggling to make the cut at that time. That there was but a single kind of common ancestor is inferred from the fact that the three emergent lineages share many genetic homologies.

To understand what this means, a helpful analogy is to imagine that you are a scholar of mythologies and you discover that a particular story—say, the slaying of a dragon by a giant Snake-God—is found in the recorded texts of three civilizations that have had no direct communication with one another. This finding would lead you to infer that the tale originated in yet an earlier civilization and was then transmitted through the three branching lines of cultural evolution. You would, in this case, expect to find embellishments specific to each culture: the Persian version, for example, might emphasize that the Snake was female, while the Ethiopian version might claim that the Snake ate the dragon's heart and became human. You might also find that a feature is present in two of the accounts and absent from the third, and this feature, you would deduce, was present in the original but dropped by one of the branches.

The lineages shown in Figure 3A have been deciphered by this same kind of reasoning. If a version of a particular gene is found in all three

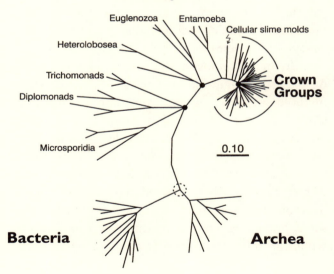

(A) **Eukaryotes**

(B)   **Crown Groups**

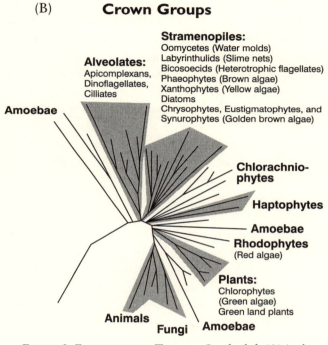

Stramenopiles:
Oomycetes (Water molds)
Labyrinthulids (Slime nets)
Bicosoecids (Heterotrophic flagellates)
Phaeophytes (Brown algae)
Xanthophytes (Yellow algae)
Diatoms
Chrysophytes, Eustigmatophytes, and
Synurophytes (Golden brown algae)

Alveolates:
Apicomplexans,
Dinoflagellates,
Cilliates

Amoebae

Chlorachnio-phytes

Haptophytes

Amoebae

Rhodophytes
(Red algae)

Plants:
Chlorophytes
(Green algae)
Green land plants

Amoebae

Animals

Fungi

∞ FIGURE 3 EVOLUTIONARY TREE ∞ On the left (A) is the "universal tree of life" with its three main branches (Bacteria, Archea, and Eukaryotes) branching off from the common ancestor (dotted circle). On the right (B) is an enlarged view of the eukaryotic "crown groups" that radiated following the Cambrian era (the "Cambrian explosion") (note that amoeba lineages have arisen several times). The scale bars do not indicate time but rather "evolutionary distance," calculated as the number of nucleotide differences found in a particular shared gene when two groups

branches, this means that the gene also resided in the genome of an ancestor common to all three groups. The lineages can then be more carefully traced by looking at the nucleotide sequences of the three versions. Recall that the information in a gene is encoded in sets of nucleotides called codons. If one group of organisms is found to have the codon CA<u>C</u> at position #43 of a shared gene, then it is judged to be more closely related to organisms with CA<u>T</u> at this position than it is to those with C<u>GA</u> or <u>T</u>A<u>T</u> at this position.

With this in mind we can move through Figure 3A. To the bottom left and right are the pathways giving rise to the present-day Bacteria and Archea. The archea are single-celled organisms that are now confined to hot sulfurous springs and other bizarre niches, but their ancestors were probably major players in earlier times when the Earth was very hot and salty. The bacteria are by far the most abundant organisms on the Earth: there are as many bacteria in your intestinal tract, or in a spadeful of soil, as there have ever been humans on the planet; indeed, there are more bacterial cells in your body than your cells! Bacteria can be said to be chemical machines: their strategy is to engage in the fastest metabolism, biosynthesis, and DNA replication possible so that they divide as rapidly as possible and generate as many of themselves as they can.

And then we wander up the long path of becoming a Eukaryote. Eukaryotes are organisms that have their DNA sequestered in a separate organelle called the nucleus (karyon). They also possess an internal cytoskeleton that allows them to move about and capture prey. Indeed, certain engulfed bacteria were domesticated and became permanent occupants of eukaryotic cells, giving rise to the modern energy-generating organelles called mitochondria and chloroplasts.

Eukaryotes apparently remained unicellular for about 2 billion years, burrowing through the soil as amoebae and swimming about in oceans and ponds as algae. Many of the modern organisms that descended from these early eukaryotes now inhabit peculiar niches—the Trichomonads, for example, are today found only in the stomachs of termites and wood-eating cockroaches. And then, some 600 million years ago, an orgy of evolutionary experimentation—the so-called Cambrian explosion—generated the numerous lineages collectively called the eukaryotic "crown groups" (upper right of Figure 3A and amplified in Figure 3B). Some of the crown-group phyla remained unicellular but became increasingly elaborate, the most spectacular being the Ciliates. Others adopted the multicellular body plan that we considered earlier, giving rise to the modern fungi, animals, plants, and various seaweeds.

If bacteria can be called chemical machines, then the eukaryotes can be called morphogenetic machines: their adaptive strategies depend more on form than on numbers. They produce fruiting bodies that rise up from the earth and allow their spores to disperse in the wind. They fashion flowers to fit the proboscis of every form of insect and, reciprocally, fashion probosci to burrow into every form of flower. They make muscles to climb and ears to hear and claws to grab. Our mantra still holds: Evolution produces organisms that can fill niches, not organisms of increasing fitness. But it can also be said that the eukaryotic groups that radiated out from the Cambrian explosion have become more complex—larger genomes, more complicated embryologies and life strategies—and hence more diverse, a pattern indicated in Figure 3B by the fanning out of the various post-Cambrian lineages.

And now a central concept. All the organisms at the tips of each radiation in Figure 3A are, evolutionarily, equally old. That is, if Figure 3A were

drawn to a true time scale and you were to trace a
path with your finger from the common ancestor to
the tip of any radiation in any of the three king-
doms, your finger would travel the same distance.
The ancient Trichomonads branched off the
eukaryotic "main line" early and spent a long time
giving rise to the creatures that now live in termite
guts; the post-Cambrian lineages branched off later
and have spent much less time reaching the twenty-
first century. But the flow of genes from the common ancestor has been constant, diverted into
countless culverts but moving steadily, from the
beginning to the present. We are all, we creatures
who are alive today, equally old, or equally recent.

Reflections

The wonders and majesty of Nature have been deep
resources for religious reflection throughout human
history. Particularly integral is the relationship
between the natural world and the peoples of
Native America, as expressed in this prayer of the
Pawnee:

> Remember, remember the circle of the sky
> the stars and the brown eagle
> the supernatural winds

breathing night and day
from the four directions.

Remember, remember the sacredness of things
running streams and dwellings
the young within the nest
a hearth for sacred fire
the holy flame of fire.

The outpouring of biological diversity calls us to marvel at its fecundity. It also calls us to stand before its presence with deep, abiding humility. Earlier we sanctified the self, and soon we will consider ways to think about our humanness with reverence and pride. But these affirmations must coexist with an understanding that all of us humans are but a tiny part of an enormous context. We are one of perhaps 30 million species on the planet today, and countless millions that have gone before. We occupy, temporally, the very last moment of the animal radiation; our species appeared only some 130,000 years ago and the cave painters 35,000 years ago. And while we animals were radiating, so too were all the other lineages of the biosphere, generating a veritable sunburst of biological ideas.

We are called to acknowledge our dependency on the web of life both for our subsistence and for countless aesthetic experiences: spring birdsong, swelling treebuds, the dizzy smell of honeysuckle. We are called to acknowledge that which we are not: we cannot survive in a deep-sea vent, or fix nitrogen, or create a forest canopy, or soar 300 feet in the air and then catch a mouse in a spectacular nosedive.

Most religious traditions ask us to bow and tremble in deference to the Divine, to walk humbly with thy God. Religious naturalism asks that we locate such feelings of deference somewhere within the Earthly whole.

Oren Lyons, Faithkeeper of the Onondaga Nation, conveyed this concept to an assembly at the United Nations:

> I do not see a delegation for the four-footed. I see no seat for the eagles. We forget and we consider ourselves superior, but we are after all a mere part of the Creation. And we must continue to understand where we are. And we stand between the mountain and the ant, somewhere and there only, as part and parcel of the Creation. It is our responsibility, since we have been given the minds to take care of these things.

VII

Awareness

A locus of human pride is our sense that we possess the capacity for a special kind of awareness, often called consciousness or self-awareness, that distinguishes us from the "dumb creatures" over which we have been assured we "have dominion."

No question, our capacity to experience awareness is in some ways distinctive, ways that we will consider later in the chapter. But no question also, these capacities are deeply homologous to the awareness inherent in all of life, and absent from nonlife. Indeed, the Earth can be wonderfully thought of as a planet shimmering with awareness. Perhaps there are other planets that so shimmer, or perhaps this is the only one. In any case, awareness is integral to life, and integral as well to our religious lives.

WHAT ARE ORGANISMS AWARE OF?

The first cells, bathed in the Eden of the primal soup, may not have been programmed to be aware. But this could not have lasted for long. Once they were forced to make ribonucleotides for themselves, for example, they would have had to locate some source of energy flow in the environment so that they could carry out this biochemistry, and throughout evolution, sets of genes have been selected that allow organisms to be aware of their circumstances and act accordingly.

Awareness is modulated by receptors and their associated signal transduction cascades. Olfactory and taste receptors detect molecular shape (Figure 2) while most other receptors respond to various forms of energy: Visual organs are stimulated by bombardments of photons; auditory organs are sensitive to the vibrations produced when air is compressed; heat receptors distinguish levels of molecular motion.

A critical focus of evolution has been the modulation of receptors such that they pick out shapes and energies of use to (or noxious to) the organism. Plants, algae, and cyanobacteria, for example, are equipped with all manner of photoreceptors that detect and absorb those wavelengths and intensities of light that are of use to their photosynthetic pathways. They also possess the means to shield themselves from light that might damage their light-harvesting systems.

More generally, much of biological evolution can be said to entail the evolution of what organisms are aware of. The first awareness systems focused on the physical and chemical properties of the planetary environment, but once a sufficient number of organisms came into existence, they became intensely aware of one another as prey or predators or symbionts. And once eukaryotic sexuality was invented, sometime around the Cambrian, countless systems were devised to recognize a mate of the correct species and the correct gender.

All sexual eukaryotes are aware of their environment, potential mates, and potential pathogens. In addition, early members of the animal radiation devised the neuron, a cell type specialized for awareness, and this made possible the avenue of awareness called consciousness.

NERVOUS SYSTEMS

Neurons all work in much the same way, from jellyfish to humans. One end of a neuron, the cell body, is equipped with receptors for a particular stimulus—an odor, for example. When odor molecules bind to their receptors and the receptors change shape (see figure 2 in chapter 3), the resultant signal transduction cascade modifies nearby ion channels such that they open up and allow a large flux of ions to enter the neuron. The influx stimulates neighboring channels to open, which in turn stimulates neighboring channels to open, with the result that an ion flux sweeps down the long length (axon) of the neuron.

When the ion flux reaches the end of the axon, the neuron is stimulated to secrete molecules called neurotransmitters onto an adjacent target cell; the junction between them is called a synapse. The target cell carries receptors for the neurotransmitters, and when the receptors bind neurotransmitters and change shape, they set off their own cascades of response. If the target cell resides in a leg muscle, the neurotransmitter/receptor interaction may cause the muscle to contract and the animal to move toward the odor-generating food. If the target cell resides in the gut, the neurotransmitter/receptor interaction may cause the gut to secrete digestive juices in anticipation of a meal.

The target cell will often be another neuron,

since neuronal cell-body membranes also carry receptors for neurotransmitters. Thus, when an olfactory neuron in a mouse's nose is stimulated by an odorant in cheese, its axon does not secrete neurotransmitters onto a leg muscle but rather onto the cell body of a second neuron, located in the brain. The brain neuron is thereby stimulated to open its ion channels, and an ion flux moves along its axon to the synapse it makes with the cell body of a third neuron, also located in the brain. This chain-reaction continues until the last neuron in the series excites the mouse's leg muscle and the mouse starts to move toward the cheese.

Once neurons are set up in such a series, their activity can be regulated. For example, the neuron that stimulates the mouse's leg muscle can be excited-in-series by odor perception in the way we have described. But its cell body is also in synaptic contact with a neuron poised to secrete an inhibitory neurotransmitter that prevents ion channels from opening. The inhibitory neuron is hooked up in series to the mouse's visual system, and if the mouse has noticed that a cat is crouching close to the cheese, the neuron will be stimulated to release its inhibitory neurotransmitter and the mouse, still sniffing, will nonetheless freeze in its tracks.

Visual perception of the cat triggers not just the "freeze" command. Nerve impulses also reach the adrenal gland, which responds by secreting the hormone adrenaline into the bloodstream. The many cascades elicited when adrenaline binds to its receptors on its target cells generate the emotional response we call fear, and within a few seconds the fear response overrides the freeze command, the leg muscles contract, and the mouse scurries back behind the refrigerator.

A key feature of animal evolution has been the increasing complexification of nervous systems. Primary receptors, like those in olfactory neurons, detect an enormous range of external stimuli, and many kinds of neurotransmitter/receptor systems then transmit and modulate responses to these stimuli. Increasingly complex networks of internal synapses have generated increasingly complex patterns of response. And primitive memory systems have given rise to increasingly impressive abilities to learn from experience and, in our lineage, to transmit these experiences as cultural understandings.

BRAINS

In advanced nervous systems, most of the neurons localize in central processing organs called brains.

The brain of every human contains about 100,000,000,000 neurons, with their axons having the collective length of several hundred thousand miles. Some of these extend out into the body, usually in series, either to take in stimuli (itch) or trigger responses (scratch). But most stay in the brain and make synaptic connections with one another. There are an estimated 100 trillion synapses in the human brain, meaning that an average cell body is in synaptic contact with 1,000 other neurons—an astonishing concept. Some of these synapses are inhibitory and others excitatory, and the target neuron proceeds to fire or not fire after integrating its various inputs. Its target, in turn, is likely to be a second brain neuron with 999 other potential synaptic influences.

Brains started out in evolution as loose collections of cell bodies called nerve rings or nerve nets, but in the higher animals they are differentiated into discrete functional domains. Figure 4 reminds us of these domains in the vertebrate lineage, where familiar anatomical units, like the olfactory bulb and the cerebellum, reappear in each brain and carry out homologous functions. The cerebral cortex increases in size during the course of vertebrate evolution and, in humans, is much larger than expected given our body size. It is in this

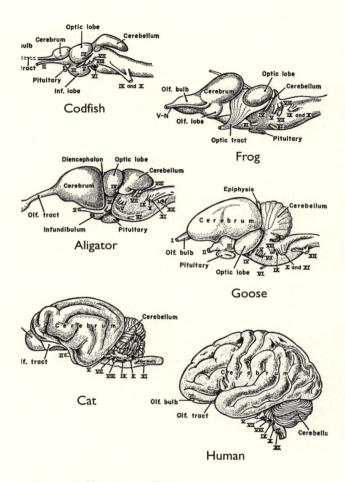

FIGURE 4 VERTEBRATE BRAINS

module that our conscious awareness is integrated.

The brains of organisms like worms are hard-wired: each member of the species has the same number of neurons in the same positions making the same synapses and performing the same functions. But as brains got larger and more complex, this stopped being the case; indeed, given that the human genome has perhaps 100,000 genes and the human brain 100,000,000,000 neurons, there's no way that each neuron could be specified by its own set of genes. Instead, the process of brain development gets set up with general instructions indicating general targets toward which particular cell lineages should move, and then the neurons engage in elaborate migrations towards these targets, with the fate of any particular cell highly dependent on the cells it moves past, the molecules they display on their surfaces, and the local nerve growth hormones they secrete. Indeed, the construction of each mammalian brain can be thought of as an evolutionary event in and of itself, with much of the fine-tuning left to contingency and selection.

Fortunately for us, most of the neural circuits that are set up during this process govern decisions that we never need to make and indeed are never even aware of. Brains process enormous amounts of physiological information to generate such results as appropriate blood pressure and breathing rate. Were it necessary to think about these matters, no poetry would ever be written. Indeed, what we consider to be our conscious selves is but a small fraction of what our brains are doing.

SELF=AWARENESS

Is an amoeba aware? Of course. But is it conscious? Is a plant conscious of light as it bends toward its source? Most of us would say no; we would reserve the term consciousness for the mental states of animals with nervous systems. But this leaves plenty of room for disagreement as to how to define consciousness. Is my cat conscious? Well, yes, I would say that she is: She takes in what she experiences and responds to it. Then how about a snail? Well, I'm not so sure. Part of the problem, of course, is that my cat seems a lot more like me than a snail, so all of my cat/snail judgments will be laden with anthropocentrism.

Granted such difficulties, I remain deeply convinced that my cat does not reflect on her self, on her own cathood, even though there is no obvious way to prove that this is so. Self-awareness is a trait that appears to have originated in the apes and has come to dominate human mentality. Awareness of our mental selves, our thoughts and feelings, awareness of awareness: We all know exactly what this means, even if none of us can describe it very well.

Neuroscience is still far from being able to explain how human self-awareness works, but some general ideas are emerging. A key concept is that we are spectators to our own awareness. We witness mental activity, represented in symbolic form, that is delivered to the "working memory" by domains obscurely called the lateral prefrontal cortex and the orbital and anterior cingulate cortices. This delivery system handles many neuronal inputs at once: inputs from long-term memory domains, emotional domains, and immediate sensory stimuli (what we are perceiving). Such inputs are "subsymbolic," meaning that we cannot decipher them consciously. They are then integrated and presented to the working memory as symbolic entities. No one yet understands what this means, nor where the working memory "is," nor what is involved in accessing these symbolic representations.

The working memory can hold about seven pieces of information at once. These can be trivial, such as my awareness of what time it is, or they can be large "chunks" of information, such as the many memories that surround the concept of my grand-mother or the many plans that constitute my sched-uale for the rest of the day. And then there is the "I" that witnesses this information, the "executive" that feels it has control over the material experi-enced. William James describes all this as well as anyone. "It is as if there were in the human con-sciousness a sense of reality, a feeling of objective presence, a perception of what we may call 'some-thing there,' more deep and more general than any of the special and particular 'senses' by which the current psychology supposes existent realities to be originally revealed."

Reflections

The fact that we take in, reflect, and encounter an I-ness is apparently at the core of religious experi-ences that we call mystical. Western traditions describe a sense of Immanence or Presence; Asian traditions variously describe states of Nirvana, of liberation, of living with the Tao. Albert Einstein underscores its centrality: "The most beautiful emotion we can experience is the mystical. It is the source of all true art and science. He to whom this emotion is a stranger, who can no longer wonder and stand rapt in awe, is as good as dead."

Throughout religious history, mystical experiences have often been interpreted as the apprehension of the Divine within or the numinous Other, and they are actively sought in prayer and ritual. In Western traditions we say that we are aware of a Spirit, that we are comprehended by something much larger, deeper, more valuable, and more enduring than ourselves and the finite universe. The encounter is inward, intensely personal, and described, if at all, with halting tongue. In Asian traditions the religious person seeks in meditation an emptying out, a receptivity, in order to experience an at-one-ness, a spiritual communion with the universe, Enlightenment.

So we raise our eyes to the heavens and we ask, Is this Other? Is this God? Is this the Perfection of Understanding? Or are these overwhelmingly powerful mental experiences, with Immanence a particularly intense form of self-awareness and Enlightenment a detachment from self-awareness so that all else can penetrate? How can we tell? And then: Does it matter?

As a non-theist, I find I can only think about these experiences as wondrous mental phenomena. But in the end it doesn't matter: All of us are transformed by their power. I have yet to reach in meditation anything approaching Nirvana, but when I am invaded by Immanence, most often in the presence of beauty or love or relief, my response is to open myself to its blessing. It is the path to the holy, taken by seekers before me and seekers to follow, and I give myself over to my mystic potential, to the possibility of becoming lost in something much larger than my daily self, the possibility of transcending my daily self. As in this hymn:

> Spirit of the living God, fall afresh on me.
> Spirit of the living God, fall afresh on me.
> Melt me, mold me, fill me, use me.
> Spirit of the living God, fall afresh on me.
>   ⚛ *Daniel Iverson, 1926*

Immanence is for me very different from cosmic Mystery. Even as I don't understand it, it is nonetheless very immediate, and experienced, and known. It becomes a part of my self that I most

cherish and value, the part that most deeply cele-
brates the fact that I am alive, the part that sustains
me through discouragement and loss.

And then, quite involuntarily, I return to the
seven pieces of information in my working memory.

VIII

Emotions and Meaning

The evolution of awareness has spun off two important capabilities.

* Organisms usually attach a *value* to the things they perceive—this is good, that is bad—which, in complex animals, is experienced via neural and hormonal emotional systems.
* Organisms usually attribute a *meaning* to something they're aware of, an ability that has for us become manifest in our capacity to think and act symbolically.

These capabilities have converged in human brains as our ability to symbolize ideas and emotions, integrate them, and present them to the working memory. We also symbolize ideas and emotions in our language and in our art, and we respond to these symbols with cognitive and emotional resonance. In parallel, we have developed the capacity to extrapolate our understanding of our-selves to the experience of other creatures. That is, we are able to experience empathy.

EVALUATION AND EMOTION

All creatures evaluate. The amoeba perceives and then moves toward a food source; it perceives and then moves away from a toxin. We move away from a dead rat; a housefly moves toward it; we move toward it, holding our breath, when our trained sensibilities remind us that we are obliged to dispose of it.

These decisions are achieved by coupling. The shape changes induced in the amoeba's food receptors are coupled, via signal transduction cas-cades, to the biochemistry of forward motility; the shape changes in the toxin receptors are coupled to backward motility. Looking at the amoeba as a whole, we can say that it possesses an affective system, an on/off, yes/no response to the stimuli it perceives.

The emotional systems associated with ner-vous systems operate in the same way, although they of course entail more complex pathways. Long before I have "thought about" whether to avoid the dead rat, I am already faint with disgust. Long before the mouse has "decided" how to respond to the cat, it has already experienced the many hard-wired responses of fear.

So what are emotional responses? The mouse's visual perception of the cat goes straight to a brain domain called the amygdala and out comes the freezing response, with no involvement of the cerebral cortex at all. And then the amygdala activates the autonomic nervous system, which stimulates the adrenal gland to release adrenaline, which causes changes in blood pressure and heart rate, again without the mouse's "knowledge." Exactly the same set of responses occurs in the frightened human: Our basic emotional reactions are ancient and hardwired survival systems that mediate our behavioral interactions with the external world.

FEELINGS

Although we have no idea what it's like to be a mouse, there is reason to believe that a mouse experiences the *emotion* fear but does not experience the *feeling* of being frightened. This is because feelings seem to arise in, and therefore presuppose, a conscious self-awareness, which is to say that a feeling is a conscious response to the unconscious fact of having had an emotional system activated.

Just as various domains of the brain somehow deliver to the working memory a set of symbols depicting how things are, so do other domains in the brain somehow deliver to the working memory information about our underlying emotional states, and it is this information that we label as an emotional feeling. Emotional information reaches the brain in subsymbolic form, often as neural and hormonal signals from our viscera and our glands. When we speak of a "gut feeling," this can be very close to the truth.

Most of us are not likely to have much difficulty dealing with this kind of analysis for a feeling such as fear—we're not very enamoured of fear in any case, and tend to think of it as a "primitive" animal instinct. But a neurobiological view of love or joy or astonishment is likely to be more troubling. Neuroscientists in fact have as yet little to tell us about love or joy or astonishment, and they are unlikely to have much to say until they understand how consciousness (self-awareness) is produced in brains. But once this is understood, then it will doubtless be the case that all feelings, including those we consider most deeply human, will be found to be created the same way that other conscious experiences come about—by establishing a mental representation of the workings of underlying processing systems.

Reductionism again. Now you scientists are turning my *feelings* into mechanisms!

Neuroscientist Antonio Damasio can help us here with his lucid version of the Mozart metaphor.

> To discover that a particular feeling depends on activity in a number of specific brain systems interacting with a number of body organs does not diminish the status of that feeling as a human phenomenon. Neither anguish nor the elation that love or art can bring about are devalued by understanding some of the myriad biological processes that make them what they are. Precisely the opposite should be true: Our sense of wonder should increase before the intricate mechanisms that make such magic possible.

But can we, nevertheless, posit that there are unique, special human emotions? After all, our capacity for language seems to have evolved only once in all of evolutionary history, so is it not possible that our brains have more to work with than just the ancient adapative emotions like fear and lust and greed and anger? Might there not be novel emotional centers, giving rise to specifically human feelings?

Perhaps. Nothing is yet known. But given Nature's track record for conservative tinkering, it seems more likely that our feelings represent elaborate combinations and syntheses of ancient emotional circuits, experienced by us in new ways. Anguish and elation may be reconfigured versions of anger and lust, without in any way being the less compelling or important.

MEANING

An amoeba carries membrane receptors for food molecules such as lactose and maltose, and it responds to these directly. But it also carries receptors for "chemoattractant" molecules that are not of interest as food but rather as cues to tell the amoeba where food is—molecules, for example, that are released when bacteria are decaying. If the amoeba is moving in one direction and its chemoattractant receptors pick up the scent at its trailing edge, this *means* that the food source is behind it, in which case the amoeba stops moving forward and starts backing up. As it moves backwards, the concentration of the cue molecules becomes greater

and greater, activating more receptors and reinforc-
ing the backwards movement until, *voilà,* it arrives
at a pile of dead bacteria and begins to feed.

Once unicellular organisms became sexual,
they acquired meaning systems that indicate that an
appropriate mate—of the correct species and opposite sex—is in the vicinity. Yeast cells, for example,
produce and secrete small pheromones that
announce their sexual identity. A cell of sex *A* has
receptors for the pheromone released by cells of sex
alpha, and when *alpha* pheromones bind to these
receptors, signal transduction cascades are set off
and genes are switched on that prepare the *A* cell to
fuse with an *alpha* partner. The *alpha* pheromone
means that an *alpha* partner is close by.

Chemoattractants and pheromones are said to
convey indexical meaning: The chemoattractants
point to (as with an index finger) the source of
food; the pheromones point to the source of mates.
Indexical meaning systems are found throughout
life: in plants, perception of red light by the seed's
phytochrome system means that the seed should
germinate; in humans, the binding of insulin to the
insulin receptors on fat-cell membranes means that
blood sugar levels are high and that glucose should
be transported into the cell interior.

Certainly the richest and most flexible meaning systems make use of symbols. A symbol, when perceived, brings something in addition to itself into awareness, such as an emotion or an idea or a cultural tradition. So far as we know, humans are unique among all creatures in their innate ability to construct and manipulate symbols—as neural representations, as languages, and as art.

So we can now come full circle and remind ourselves of what we know full well: Our working memory is not so much presented with stand-alone "thoughts" and stand-alone "feelings" as with intricate integrations of the two. My grandmother chunk comes into my working memory and as I experience her smile and her stride, I also experience my admiration for her energy and my impatience with her fussiness and my deep sorrow that she is no longer alive. The admiration and the impatience may come to me as remembered feelings that I don't actively experience, but the sadness wells up in my body as a primal heaviness, and infuses her smiling image. The image of my grandmother has become a complex symbol of cognitive and emotional meanings.

Reflections

How are emotion and feeling and meaning and symbolism brought into a religious framework? Let's begin with William James. "There must be something solemn, serious, and tender about any attitude that we denote religious. If glad, it must not grin or snicker; if sad, it must not scream or curse."

A recent visit to the Brooklyn Museum of Art validated this insight. In collection after collection, I found I had no difficulty figuring out which pieces were considered "religious" and which "secular," even when the works came from cultures I knew nothing about. The religious pieces were infused with feeling and immediacy, but always in the context of a solemnity or tenderness. The masks, however frightening, were also noble; the fertility figures were dignified; the totems and sarcophagi were carved with deep respect for the beliefs they symbolized.

And then there was the symbolism itself. Each piece was redolent with meaning, and each conveyed complex feelings, like hope or assent. A powerful metaphor demonstrates the depth of one's understanding, and this is what religious art has always been about. Each totem elicits the religious feelings embodied in the tradition and made manifest by the artisans. Each crucifix calls us to the pathos of Christ. Each image of the Buddha invokes a reflective serenity.

It is, I believe, our capacity to apprehend the meaning and the emotion embedded in symbols that endows us with our capacity for empathy. Empathy, in the end, is the ability to imagine what it is like to experience something, the ability to put oneself in another's shoes. It is a higher-primate capacity but not exclusively human. A young bonobo chimp was observed to injure a bird while playing with it. She picked up the comatose body, carried it with her to the top of a tree, and opened out her hand. When the bird still lay there limply, she wrapped her legs around the tree trunk, held the bird by its wings, and then opened and closed the wings several times, apparently trying to help it start to fly.

Once there is empathy, then there can be the feeling we call compassion. A version of the Golden Rule—Do unto others as you would have them do unto you—is found in most religious traditions. It is as we can imagine being the least of these that we can begin to experience the anguish of deep poverty or deprivation. It is as we are able to identify with the oil-soaked shore bird and the bewildered moose that they come to symbolize our environmental concerns. And emergent from our sense of compassion, in mortal conflict with our insistent sense that we should win, is our haunting sense that things should be fair.

IX.

Sex

Eukaryotic sex is both ancient and ubiquitous: It arose some time prior to the Cambrian explosion, and is found in all the phyla that trace back to the Cambrian (Figure 3).

Sex necessitates the coming together of two genomes, and thus it necessitates the finding of a mate. Therefore, the origin of sex marks the onset of biological relationship—as contrasted with the solitary asexual existence of the bacteria and amoebae—a theme developed in the next chapter. Once procreation was handed over to germ cells and embryos and offspring, moreover, their protection assumed vital importance and, in animals, was entrusted to strong emotional instincts. These we consider later in this chapter.

But first we have to look at what sex is about.

WHAT DOES SEX ENTAIL?

A genome contains all the genetic information needed to make an organism, but in eukaryotes it is not encoded in a single piece of DNA. Instead, it is divided up into a number of lengths of DNA called chromosomes. A useful analogy here is to an encyclopedia. All the information in an encyclopedia could be printed in a single huge volume that is rolled about in a wheelbarrow, but it is more manageably organized as a set of volumes, the first containing all the A-B information, the second the C information, for a total of, say, nineteen volumes. A chromosome is equivalent to a volume, and the full set of nineteen chromosomes comprises a genome. The genome of each species is apportioned to distinctive numbers of chromosomes: cats happen to have nineteen, while humans have twenty-three, goldfish fifty, flies four, corn ten.

The chromosomes reside in the nucleus of a eukaryotic cell, and we have already considered how their component genes come to give rise to organisms. Now our concern is how they are transmitted sexually from parent to offspring.

Sex entails making two kinds of cells: a haploid cell with a single complete set of chromosomes, and a diploid cell with two complete sets of chromosomes. The diploid cell, not surprisingly, is formed when two haploid cells fuse together. Reciprocally, haploid cells arise when diploid cells engage in a halving of their chromosome number. We can look at each process in turn.

Forming Diploid Cells

The formation of a diploid cell occurs during fertilization: Two haploid gametes, a sperm or pollen grain from the male and an egg from the female, fuse to form a single diploid cell called a zygote. Returning to the encyclopedia analogy, if we could color all nineteen volumes from the sperm blue and all nineteen volumes from the egg pink, the zygote would have a complete set of blue volumes and a complete set of pink, thirty-eight in all, such that every entry, every gene, is present twice.

There are important advantages to having two sets of instructions. Let's say that there's a large printer's error in the "Calcium Ion Channel" entry in the blue volume 2. Chances are that the pink volume 2, printed in a different year in a different city by a different typesetter, won't have this same error (although it may well have other errors in other entries). If you have both volumes, you have access to readable information from the pink version and hence can construct a serviceable ion channel. Expressing this in genetic terms, we say that diploid cells, carrying two sets of instructions for making each kind of protein, are less vulnerable to deleterious mutations than haploid cells are.

Forming Haploid Cells

Now we can look at the reciprocal process, the formation of haploid cells from diploids. In Nature this happens in many contexts; we can consider here how it occurs in the testis of a cat.

The task is to generate haploid sperm with nineteen chromosomes from diploid precursor cells with thirty-eight chromosomes, nineteen of which are pink (from mother) and nineteen blue (from father). If this were accomplished by grabbing nineteen chromosomes at random, the result would almost certainly be a disaster: The sperm would wind up carrying, say, a pink and a blue chromosome #1, no chromosome #2, a pink chromosome #3, no chromosome #4. That is, you'd almost certainly create a sperm with an incomplete set of instructions for making a cat.

Therefore, the rule is that you have to include a complete set of chromosomes—a complete set of encyclopedia volumes—in every haploid sperm nucleus. But this doesn't mean that they have to all be pink or all be blue. You can have a pink #1 and #2, a blue #3, a pink #4, and a blue #5, so long as you wind up with one of each kind, a complete set.

All this takes place during an elegant cellular process called meiosis, during which the chromosomes are carefully segregated and assorted such that complete sets are generated. And the consequences are profound. Our original diploid male cat had nineteen pink chromosomes and nineteen blue, whereas each sperm he produces will contain a distinctive full collection of nineteen chromosomes, some pink and some blue. When any of his sperm manages to fertilize an egg, the egg nucleus will contain a set of nineteen chromosomes that has also been shuffled by meiosis. Therefore, while the resultant diploid zygote will have thirty-eight chromosomes, two complete sets of genetic information, these will be very different from the sets that were present in the parents.

Another way of stating this is to say that although parents each contribute half of their genetic endowment to a child, they basically end up with a stranger.

WHAT DOES SEX ACCOMPLISH?

If we look at the female cat who generated the egg just fertilized, her chromosomes (we can call nineteen of them orange and nineteen of them purple) are expected to carry a different spectrum of mutations from the pink/blue set contributed by the male. Specifically, the calcium-channel gene in the orange version of her chromosome #2 might specify a channel that transports ions particularly rapidly, whereas her purple chromosome #2 might carry a slow channel gene. If a fertilized egg winds up with an orange and a pink chromosome #2, this means that the kitten will have two kinds of channels: one rapid (orange) and one normal (pink). A littermate might inherit orange and blue versions of the gene and therefore only transport calcium rapidly—the blue version is dysfunctional. In genetic terminology, we say that each zygote, and hence each kitten, carries two alleles of the channel gene, with allele meaning "a different version."

Now we can go to all cats, and look at every calcium channel gene in every chromosome #2 in the entire species. We might in this case find twelve alleles of the gene; we can drop our color-coding analogy and call them C_1–C_{12}. Some will encode nonfunctional proteins, like our blue example, with deleterious mutations at various positions in the gene sequence. Others will carry codon changes that make no difference to the rate of calcium flux—so-called neutral alleles. Others will specify channels that work at various rates. The survey will also reveal that some of the alleles are more abundant than others. 62% of the genes might be C_2, 13% C_3, 3% C_{12}, and 0.1% C_1. But they'd all be represented in the cat "gene pool."

If we move along chromosome #2, we come to the next gene—let's say it codes for an enzyme involved in making black fur. Looking at all cats, we find that 44% have the allele B_1, which results in jet black hair, 17% have allele B_2, which yields charcoal gray, and 39% have B_3–B_{16}, all of which encode dysfunctional enzymes so the hair is white. The next gene I, and its series of alleles, may be involved with implantation of the fetus, the next P with purring.

So, virtually every cat chromosome #2 will be different from every other. The first might read C_1,B_3, I_6, P_3, the second C_1, B_7, I_6, P_1, and so on. The same, of course, will also be true of the other eighteen kinds of chromosomes. Since the cat genome contains perhaps 90,000 genes, each chromosome will on average carry some 4,500 genes, meaning that the cat species harbors a huge chromosome diversity.

Running through these concepts one more time with the encyclopedia analogy, we can imagine thumbing through countless sets of nineteen-volume encyclopedias, with each volume having on average 4,500 entries. The first entry in volume 1 is always Aardvark, but there may be eighteen versions (alleles) of the Aardvark spiel in the world-wide volume-1 pool, some informative, some mediocre, and some unintelligible. The second entry is always Aaron, with twenty-seven versions. If you pick up a volume 1 in one bookstore you might find Aardvark 17 followed by Aaron 3; the next bookstore might have Aardvark 9 followed by Aaron 14.

EVOLUTIONARY STRATEGY

And now, finally, we can put all of this together. There are nineteen different kinds of cat chromosomes, each one of its kind carrying a unique lineup of alleles. If we think of the whole pool of cat chromosomes on the planet as a gigantic collection of playing cards, then in one generation the cards are dealt out as sets of nineteen cards held in countless gamete-nucleus hands. Each fertilization brings two sets together and creates a diploid kitten that, when it matures and makes sperm or eggs, shuffles the two sets together and deals out new haploid hands. Any one of these then combines with a second new hand to make a new diploid kitten, with many such fertilizations creating the next cat generation.

What this means is that each sexual generation is, in effect, a whole new card game, with each cat holding a unique diploid hand and the entire gene pool dealt into countless diploid hands, all subject to natural selection. The hands—and hence the cards—that survive are shuffled and dealt again, with the next generation of diploid hands again substrates for natural selection.

This strategy is beautifully designed for generating different kinds of organisms. Each diploid hand is in effect a new experiment in making a cat. A given allele is placed in a nucleus with 179,999 other genes, many of which it has probably never coexisted with before, and even subtle differences in the time of appearance, shape, or stability of the resultant protein products may generate subtle differences in the cat's ability to hunt, resist disease, or produce offspring. The next diploid, with a different palette of alleles, will come up with a slightly different version of a cat.

In effect, sexual populations deal all their hands, strut their stuff, at every generation, rather than going out on a limb and specializing in one particular phenotype the way an asexual population tends to do. Specialization can definitely be a good idea over the short haul, when a particular facet of a niche can be exploited by a particular kind of creature. But it is vulnerable to the fact that most niches keep changing.

Overall, then, two reproductive strategies seem to win the evolutionary lottery every time. The first is to be asexual and make as many specialized organisms as you can before the niche changes—the strategy of the bacteria. The second is to be sexual and make enough different kinds of organisms in one generation that at least some survive the vagaries of the niche and make enough different kinds of new organisms that the whole enterprise keeps going.

NURTURE

The advent of sex marks a whole new idea in the history of organisms. While the overall goal—the transmission of genomes from one generation to the next—is the same as with asexual organisms, the genomes are now entrusted to a new class of individuals, the immature offspring. Therefore, the nurture of offspring is fully as important as surviving long enough to produce them.

Nurture is manifested in countless ways. Plants go to great lengths to ensure that their fertilized ovules are surrounded with hardy seed coats and fruity tissues. Butterfly larvae snuggle in cocoons; the social insects stagger out of disturbed nests with larvae in their mouths to carry to the next refuge. And the vertebrates, particularly the mammals and birds, have devised a stunning array of behaviors to assure the survival and maturation of their progeny.

Reflections

We have considered thus far two ways to think about caring. We have acknowledged our deep genetic homology with all of life and the affinity, the fellowship that emerges from that acknowledgment. We have also celebrated our capacity to experience empathy with other creatures and respond to their concerns as our own. And now we encounter our biological imperative to nurture our offspring, sacrificing, if need be, our lives on their behalf.

My own experience with this imperative came when, alone on a beach with my youngest son, I saw him being dragged out to sea. I jumped in, fully clothed, and as I swam out I realized that I was taking in huge amounts of water as I navigated the strangely turbulent surf. My brain displayed the headline: This Is How People Drown in Rip Tides. I looked ahead at James's terrified face bobbing above the waves, and the next realization came to me not as a headline but as an understanding: Either both of us survive or both of us drown. I reached him and pulled him to shore with a calm conviction that was somehow outside myself, and as we stood together on the empty beach, I absorbed my new self-knowledge: I am endowed with an inherent maternal altruism, unrehearsed, that is poised to flood my being whenever my children are in danger. There is no way to describe the joy that attends this kind of knowledge.

It seems likely that the emotional circuits invoked when we contemplate our deep evolutionary affinity with other creatures, and when we are infused with compassion, will turn out to map closely onto the circuits that drive our parental instincts, emotions that generate such feelings as tenderness and warmth and protectiveness. These same emotions extend to our understanding that the Earth must be nurtured, an understanding embedded in many religious traditions.

> There are creatures whose children float away
> at birth, and those who throat-feed their young
> for weeks and never see them again. My daughter is
> free and she is in me—no, my love
> of her is in me, moving in my heart,

changing chambers, like something poured
from hand to hand, to be weighed and then
  reweighed.

  ✤ *Sharon Olds, 1996*

We nurture our children selflessly. But we also
recognize them as our most tangible sources of
renewal—for a child, the world is always new.
Renewal has been a religious theme throughout the
ages, be it the Jews exhorted by Isaiah to return to
Jerusalem after their exile in Babylon or the disci-
ples exhorted by Jesus to seek the redemption of the
spirit. Theists find that they can renew their personal sense of worth through petitions to God for
atonement and grace. All of us see in children—our
own and all children—the hope and promise of
what we humans can become. As the forbears of
our children we are called to transmit to them a
joyous and sustainable vision of their future—
meaning that we are each called to develop such a
vision.

X.

Sexuality

Sex, as we have seen, has everything to do with the adaptive strategy of eukaryotic organisms, and it generates the seminal necessity to nurture offspring. But sex generates another important consequence as well: Gametes carrying haploid genomes must, at each generation, fertilize other gametes to create new diploid organisms.

Bacteria and amoebae have no such burden; their sole obligation is to go through a cell cycle, divide in two, go through another cell cycle, divide in two again, ad infinitum. Whereas sexual creatures, at a minimum, must produce gametes that find, recognize, and then fuse with gametes of the same species and opposite gender, a far more ambitious proposition. Still more ambitious are the animals that keep their gametes inside their bodies rather than spewing them out into the water or the air. In these cases, it becomes necessary to first identify the animal of the correct species and opposite gender, and then engage in copulation with that animal such that the gametes can fuse.

These strategies entail relationship, if only brief and reflexive, between sexually mature males and females, and they are, of course, antecedent to the elaborate emotional networks that govern human sexual relationships.

ATTRACTION

All organisms locate their mates, or the gametes of their mates, by some form of sexual attraction. Attraction can entail a simple receptor-mediated interaction, like the binding of the *alpha* pheromone to its receptor on the yeast A cell, or the binding of a protein on the surface of a starfish sperm to its receptor on the surface of a starfish egg. But even such simplicity can become complex, as when higher plants produce elaborate flowers so that insects or birds will brush against their anthers and transport their pollen to the stigmas of other flowers. And then, once the pollen binds to its correct receptor, it is programmed to send a long pollen tube down into the ovary where it locates and fuses with a receptive ovule.

Animals with nervous systems take the behavioral possibilities for sexual attraction to every possible limit. Fireflies pulse, houseflies beat their wings, moths send out musk, fish dance, frogs croon, birds display feathers and song, mammals strut and preen. If this is a planet shimmering with awareness, then a great deal of that awareness is focused on the sexual signals that creatures send to one another.

If we look to our closest relatives, the bonobos and the chimpanzees, we find quite different approaches to sexuality. The bonobos have sex with one another—male with male, male with female, and female with female—about ten times a day, often to reduce levels of conflict or solidify alliances, but often just because they seem to enjoy it. The chimps, in contrast, have only heterosexual sex, and only when the female is in heat. Neither group is monogamous, and although dominant chimp males attempt to monopolize one or several females, "dalliances" occur frequently.

The range of human sexual behaviour includes all of the above. In addition, humans profess allegiance to the concept, if not always the practice, of committed marriage. This commitment feeds into the second facet of sexuality, the need for other.

THE NEED FOR OTHER
All sexual organisms need to attract a mate if they are to transmit their genes to the next generation. Whether this need for sexual relationship is experi-

enced consciously takes us back to the issue of which creatures are conscious, an issue we can bypass by agreeing that whether or not humans are unique in experiencing this need, what we experience is an awareness of emotional pathways that have deep evolutionary roots.

Humans also rely on one another for the nurture and care of their dependent offspring. Much of this has come to be accomplished by larger social groups—clusters of males hunting for game and defending against predators, clusters of females gathering fruits and tending the hearth. But human pair-bonding is encountered in all cultures and appears to be instinctive, at least while the children are young.

The instinct to engage a mate to help with child-rearing is accompanied by the reciprocal instinct in children (and in all young mammals and birds) to form strong relationships with their all-important parents. Again we do not know whether the need for parenting penetrates conscious awareness in other young animals, but it seems probable that our affection for our parents flows through emotional networks that establish parent–offspring bonds in other mammals.

Psychologists have long posited that our love/need for our parents emanates from the same impulses as those that drive our love/need for our mates, even though they are expressed at different stages of maturity and experienced as very different sets of feelings. Certainly they have in common an extraordinary intensity. At least at the outset, our emotional responses to our parents and to our mates are thoroughly wondrous, thoroughly compelling, and deeply joyous.

Alas, of course, intimate relationship is inherently fraught with conflict. We confront, often clumsily, the imperative that we separate from our parents while retaining affection for them. We struggle to accommodate our love for our mates, and our need for their reliability and trust, with the experience of what we elect to call temptation and lust. When we find ourselves estranged from our mates, we are torn apart by jealousy, loneliness, desolation, and anger. We fear disapproval and abandonment. We can become deeply confused about our sexuality. It is all very complicated.

Reflections

Given the complexities of human relationship, an enormous attraction of the monotheistic religions—Judaism, Christianity, and Islam—is that they offer the opportunity for intimate relationship with a deity. Indeed, they suggest that the most stable and fruitful outlet for passion and dependency is in relationships with the Divine.

Judaism initiated such a path with its concept of a Father God, a stern and judgmental father to be sure, but one who can also offer protection and even affection, as in Psalm 23:

> The Lord is my shepherd; I shall not want.
> He maketh me to lie down in green pastures.
> He leadeth me beside the still waters.
> He restoreth my soul.
> He leadeth me in the paths of righteousness
> for his name's sake.
> Yea, though I walk through the valley of the
> shadow of death,
> I will fear no evil, for thou art with me;
> Thy rod and thy staff they comfort me.
> Thou preparest a table before me in the presence of mine enemies.
> Thou anointest my head with oil; my cup runneth over.
> Surely goodness and mercy shall follow me all
> the days of my life,
> And I will dwell in the house of the Lord for
> ever.

And then Christianity (and Islam) took this all

the way. Christian doctrine certainly implores us to feel compassion for others, but it speaks with particular poignancy to our longing for relationship. The reward of Christian faith, we learn, is the inexhaustible, unconditional love that flows from God the Father and Mary the Mother and Christ the Redeemer. They are there *for* us, they listen and respond, they will never abandon us, and seek only our love in return. As in these hymns and prayers.

> Jesus, the very thought of thee with sweetness
>   fills the breast;
> But sweeter far thy face to see, and in thy
>   presence rest.
> O hope of every contrite heart, O joy of all
>   the meek,
> To those who fall, how kind thou art! How
>   good to those who seek!
> But what to those who find? Ah, this nor
>   tongue nor pen can show;
> The love of Jesus what it is, none but his loved
>   ones know.
>   ❦ *Bernard of Clairvaux, 1153*

> Jesus, priceless treasure, source of purest plea-
>   sure,
> Truest friend to me,

Long my heart hath panted, till it well-nigh
    fainted,
Thirsting after thee.
Thine I am, O spotless Lamb, I will suffer
    naught to hide thee,
Ask for naught beside thee.

                        ❧ Johann Frank, 1653

Jesus, lover of my soul, let me to thy bosom
    fly,
While the nearer waters roll, while the tem-
    pest still is high.
Hide me, O my Savior hide, till the storm of
    life is past;
Other refuge I have none, hangs my helpless
    soul on thee;
Leave, ah! leave me not alone, still support
    and comfort me.
All my trust on thee is stayed, all my help
    from thee I bring.

                    ❧ Charles Wesley, 1740

Softly and tenderly Jesus is calling, calling for
    you and for me;
See, on the portals he's waiting and watching,
Watching for you and for me.
Come home, come home; you who are weary,
    come home;

Earnestly, tenderly, Jesus is calling, calling, O
sinner, come home.

         ✄ *Will L. Thompson, 1880*

So we arrive here at what is, for many, the
heart of it all. If there is a major tension between an
approach like religious naturalism and the
monotheistic traditions, it centers on the question
of whether or not one believes in a personal god.
Most people raised in the context of theistic traditions would probably say that "being religious"
means "believing in God." Indeed, when reminded
that personal gods are not inherent in such systems
as Buddhism or Taoism, they would likely question
whether these traditions are really religions and not
something else, like philosophies.

For me, and probably for all of us, the concept
of a personal, interested god can be appealing,
often deeply so. In times of sorrow or despair, I
often wonder what it would be like to be able to
pray to God or Allah or Jehovah or Mary and
believe that I was heard, believe that my petition
might be answered. When I sing the hymns of faith
in Jesus' love, I am drawn by their intimacy, their
allure, their poetry. But in the end, such faith is simply not available to me. I can't do it. I lack the
resources to render my capacity for love and my
need to be loved to supernatural Beings. And so I

have no choice but to pour these capacities and needs into earthly relationships, fragile and mortal and difficult as they often are.

Theism *versus* Non-Theism. The choice has been presented to us as saved *versus* damned, holy *versus* heathen. But when I talk to thoughtful theists, I encounter not a polarity but a spectrum. Belief and faith in supernatural Being(s), when deeply wrought, are as intensely personal and individual and dynamic as our earthly relationships. They add another dimension, another opportunity for relationship, to be sure. But those of us incapable of embracing that dimension remain flooded with opportunities to open ourselves to human relationship and hence to fill our lives with the religious experience of love.

What the monotheistic traditions offer to all of us, theists and non-theists alike, are challenging and enchanting images and evocations for how to best love. Michelangelo, unsurpassed in his ability to render in visual art the spirit of Christian love, writes of his earthly passions with the same imperative.

> With your beautiful eyes, I see a gentle light
> my blind ones could never see;
> On your feet, I bear a burden
> my lame ones could never bear.

With your wings, I fly though featherless;
By your mind I'm lifted ever upward;
At your whim, I pale or blush,
    cold in the sun, warm in the cold of winter.
In your desire alone is my desire;
    my thoughts are forged in your heart,
    my works are breathed in your breath.
Alone, I am like the moon, itself alone;
    our eyes see it in the heavens
    only as the sun enlightens it.

                      ❧ *Michelangelo, 1534*

$\mathcal{X}I$

Multicellularity and Death

THE GERM/SOMA DICHOTOMY

Many kinds of sexual algae and fungi are single-celled. Each cell/organism is either a haploid male or a haploid female, and each has two options: it can replicate and divide and replicate and divide to generate millions of identical copies (a mitotic clone) of itself, or else it can recognize and fuse with a cell of the opposite sex to produce a diploid zygote. The zygote switches on genetic programs that allow it to form a protective spore coat around itself and go into dormancy. And then, when circumstances are favorable, the spore undergoes meiosis and releases haploid male and female organisms that are pink/blue mixtures of their parents, and these again either cycle mitotically or else mate with one another.

Multicellular eukaryotes evolved from such single-celled creatures at the time of the Cambrian.

We have already considered how multicellular organisms produce all manner of specializations by expressing different sets of genes in different sets of cells. Omitted from that account was the important fact that all multicellular organisms are sexual. Indeed, the invention of sex was necessary for multicellularity to evolve.

To understand what this means, we can consider the diploid zygote—the fertilized egg—of a multicellular animal. Whereas the algal zygote has but modest potential—it can form a spore coat and it can undergo meiosis—the animal zygote proceeds to cleave into two cells, and then four and then eight, with each cleavage generating daughter cells that remain together as a developing embryo. And then all of them start to specialize.

As we have said, each cell expresses only a subset of the genes it possesses, a differential that plays itself out in space and time. Let's focus on one of the cells in an 8-cell embryo, a cell programmed to switch on a certain set of genes. In the 16-cell embryo this cell has given rise to 2 daughter cells, both containing the protein products of these genes, and the products cause a second subset of genes to switch on. In the 32-cell embryo, the products of the second subset initiate a signal-transduction cascade that induces the now-4 daughter cells in the lineage to move together to a new location and, several cleavages later, to move into the interior of the embryo in a process called gastrulation. Following gastrulation, the lineage (now 512 daughter cells) is subject to several fates: 64 of the cells at one end of the embryo activate a set of genes that allow their daughters to differentiate into gut cells; another 8 near the midline activate the program that ultimately generates the heart; and so on.

Early in this process of embryogenesis, certain cells switch on sets of genes that commit them to become germ-line cells—precursors of the egg or sperm cells that are uniquely capable of undergoing meiosis. These migrate into what will become the animal's gonads, where they remain dormant until sexual maturity and then begin undergoing meiosis to produce haploid gametes.

We can now appreciate the beauty of this arrangement. The dichotomy between the germ-line cells and the remaining somatic cells effectively parcels out the job of being alive. Transmission of the genome to the next generation is entrusted to the germ line, while negotiating the niche so that the germ cells are successfully transmitted is entrusted to the soma. The germ line is safely sequestered in gonads, nurtured by surrounding tissues, its genomes released only at appropriate times; the somatic cells are the ones that perceive and move and sprout feathers and pump blood and make love.

MORTALITY AND IMMORTALITY

One of the fates that is often programmed into a cell lineage during the course of embryogenesis is that those cells should die. Thus, the limbs of a human embryo initially terminate as blunt stubs, after which sets of cells die in order to create separate fingers and toes. And every autumn, in every deciduous tree, the cells at the base of each leaf stem are programmed to die such that the flow of nutrients is cut off and the leaves themselves die.

The more general fate of the soma is that the whole soma dies. If this death is premature, before the germ line has had time to be successfully transmitted to the next generation, we say that that organism was either unfit (an insect incapable of flight) or unlucky (an insect eaten by a bird). But if it happens after the germ line has successfully participated in the production of sons and daughters, then we say that the organism has served its biological purpose. Natural death may occur after only a few days of life, as with some kinds of adult insects, or it may be postponed for hundreds of years and hundreds of attempted procreation cycles, as is the case for some kinds of trees.

Eventually, though, the sequoias die just like the dragonflies. If we don't die by accident or infection or because of the failure of a particular organ, we die because we just get old. A friend describes her husband's last two years before his death at the age of ninety-one: "It just got slower and slower, and less and less, and then he stopped being interested in eating, and then in drinking, and then he stopped breathing."

So is there such a thing as an immortal organism? The answer is yes, but immortal organisms are by definition very limited in complexity. For example, there is no death programmed into the life cycle of a bacterium or an amoeba. For sure, the cells can be killed by boiling or starvation—the individuals are fully mortal—but they don't *have* to die. The same is true for the sexual single-celled algae that we grow in my laboratory. The cells need to have sex when they are in the wild—they must form heavy-walled zygotic spores to protect their genomes from freezing and dessication—but under our care they will keep on dividing indefinitely by mitosis as long as we provide them with light and nitrogen salts. By the same token, tumor cells, in scientific terminology, are said to be "immortalized." They carry somatic mutations in key cell-cycle genes such that they don't know when to stop dividing, either in our bodies or in the laboratory.

But once you have a life cycle with a germ line and a soma, then immortality is handed over to the germ line. This liberates the soma from any obligation to generate gametes, and allows it to focus instead on strategies for getting the gametes transmitted. And since morphogenesis is the key niche-negotiating strategy of eukaryotes, multicellular eukaryotes, freed of constraints, have generated every complex morphological structure imaginable: wings, gills, eyes, leaves, glands, claws, bark, nostrils, tentacles. All of these parts are highly specialized, and although each cell in each part retains two full copies of the genome, transmission of these somatic genomes to the next generation is not included in the arrangement. The arrangement is that the parts will do their utmost to ensure the transmission, and often the nurture, of the germ line, and then they die.

One of these "parts" is my brain, the locus of my self-awareness. My brain developed with nary a backward look at gene transmission or immortality. The whole point was to make synapses, strengthen them, modulate them, reconfigure them, with countless neurons dying in the process and countless more dying during my lifetime, many as I sit here typing. It is because these cells were not committed to the future that they could specialize and cooperate in the construction of this most extraordinary, and most here-and-now, center of my perception and feelings.

So our brains, and hence our minds, are destined to die with the rest of the soma. And it is here that we arrive at one of the central ironies of human existence. Which is that our sentient brains are uniquely capable of experiencing deep regret and sorrow and fear at the prospect of our own death, yet it was the invention of death, the invention of the germ/soma dichotomy, that made possible the existence of our brains.

Reflections

All religions offer us a way to think about death, usually in the context of some form of immortality. We know about the heaven/hell of Western traditions and the reincarnation cycles of Asian traditions, but in fact the concept of immortality is global. The Bwende, in the Congo, carved icons to the Four Moments of the Sun: Dawn (the beginning of life), Noon (life at its fullest), Sunset (the end of life's journey), and a Second Dawn (for those who have lived an exemplary life). The Egyptians developed an elaborate Afterlife ruled by King Osiris and inhabited by numerous gods. The Taoists look to Fei-sheng, the ascension to heaven in daylight. The Muslims anticipate resurrection (yaum al-qiyama) and final judgment (yaum al-din).

Religious naturalism offers two responses to human death. The first is the response to the death of someone loved, or a death that is premature or senseless. These directly ravage our personal fabric of relationship, or activate our empathy and compassion, and we experience unmitigated loss and grief. I was told of a school-age child whose mother was killed in an automobile accident—how weeks later he would go into her clothes closet and bury his face in her dresses so he could smell her smell. I am undone by his savage loss, and outraged by her death, even though these people are strangers to me. Our sorrow at the death of others is a universal human emotion that transcends cultural and religious particularities. Indeed, ape mothers have been observed carrying their dead babies around for several days, suggesting that this form of grieving far antedates our humanness.

And then there is the response to the fact of death itself, and, in particular, to the fact of my own inevitable death. When I wonder what it will feel like to be dead, I tell myself that it will be like before I was born, an understanding that has helped me to cope with my fear of *being* dead. But what about the fact that I will die? Does death have any meaning?

Well, yes, it does. Sex without death gets you single-celled algae and fungi; sex with a mortal soma gets you the rest of the eukaryotic creatures. Death is the price paid to have trees and clams and birds and grasshoppers, and death is the price paid to have human consciousness, to be aware of all that shimmering awareness and all that love.

My somatic life is the wondrous gift wrought by my forthcoming death.

XII

Speciation

Eukaryotic sex has given rise to the evolution of nurturing, the evolution of love, and the evolution of multicellularity and death. Here we consider a final manifestation: Sexual eukaryotes came to adopt the evolutionary pattern known as speciation, which segregates organisms into those that will and will not mate with one another. Such segregation allows each species to develop distinctive traits, and has come to generate much of our biodiversity.

THE DYNAMICS OF SPECIATION

Meiosis is a fair lottery, providing each allele with the chance to be transmitted to the next generation and expressed in concert with the other genes that find their way into the same zygotic nucleus. Natural selection then acts on the particular combinations that are generated, and the surviving alleles get shuffled and re-dealt to new zygotic nuclei.

This arrangement, we have said, allows a species to display its full range of variation at each generation. To be sure, certain alleles—certain versions of gene sequences—come to be more prevalent under certain conditions, but this can be rapidly reversed should conditions change, a famous example being the dark moths that prevailed over light ones when pollutants killed the light-colored lichens that covered the dark bark, and the return of the light moths once pollution came under control and the lichens re-established themselves. Both the light and the dark alleles were harbored in the moth population; natural selection—in this case, bird predation—influenced their relative frequencies.

Such changes in the frequency of alleles within a species can be contrasted with a second phenomenon, the origin of a new species. Although the dynamics are not yet understood, the outcome is well known: Members of a new species fail to generate fertile offspring when placed in contact with members of their parental species or other sister species. This can happen because sexual behaviors have changed—the male insect's mating dance may no longer elicit receptivity in the female—or because the eggs can no longer be fertilized by the sperm, or because the embryos fail to develop properly and die, or because the adults are for whatever reason sterile, an example here being the mule.

The consequences of speciation can be described in the context of our playing-card analogy. Speciation creates a new deck of shuffling genomes. If the new deck has only recently become isolated from the old (parental-species) deck, the two will share a great many alleles. But because they are not being shuffled together and are subject to different kinds of natural selection, the two decks will come to have quite different frequencies of alleles. Importantly, the new deck will also come to contain new cards—new genetic ideas—that endow the new species with distinctive sexual and adaptive traits not present in the parental species.

Figure 5 illustrates the pattern of speciation and, hence, the pattern of eukaryotic evolution. Each species is represented by a teardrop-shaped entity. The length of the teardrop indicates the length of time a species exists before it goes extinct. The width of the teardrop indicates the "niche dimensions" of the species—the range of the global habitat that can be successfully negotiated by the members of the group at each generation. Each nar-

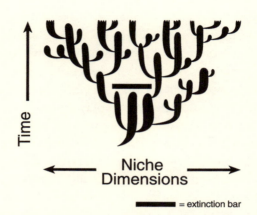

Time

← Niche →
Dimensions

▬▬▬ = extinction bar

∞ FIGURE 5 SPECIATION PATTERNS ∞ The "tree" represents a sexual lineage (also called a clade) moving through time (vertical) and expanding its niche dimensions (horizontal). Each teardrop-shaped entity is a species, originating at its narrow base and going extinct at its rounded tip. The bar represents an abrupt disruption of the niche such that the three species inhabiting that dimension go extinct.

row sideways branchpoint represents a speciation event, wherein a new species not only becomes sexually isolated but also comes to occupy a distinctive niche dimension. The "extinction bar" in the figure represents a catastrophic change in the niche that wipes out certain species, but the lineage as a whole keeps going and indeed, with time, the devastated niche becomes repopulated with new species.

A species, then, is characterized both by its

macro-distinctiveness and by its micro-diversity. The ornithologist can recognize which finch species is which by sets of distinctive field markings, many of which are used by the finches themselves to select a correct mate. The ornithologist can also describe in detail the nesting behavior, preferred diet, and flight patterns of each species. But at the genetic level, each finch in a species carries but a subset of the total gene pool, meaning that each bird has its own set of nesting-eating-flying characteristics, even though they conform in general to the distinctive patterns of the whole. A species can therefore be said to possess buffered distinctiveness.

EVOLUTIONARY TIME

The narrow-necked speciation events diagrammed in Figure 5 happen much more rapidly than the length of time a species survives, which is to say that speciation does not occur by some gradual divergence within an existing species, but rather as an abrupt branching-off. "Abrupt" can translate into thousands of years, and a species typically persists for some 10 million years, so these temporal concepts need to be thought of in relative terms.

This is a good place to pause and ask what it means to talk about thousands or millions of years, concepts that are hard to grasp because our experience with time is so limited. When we say that apes and humans diverged recently, only 5 million years ago, we are saying two things: that the divergence was indeed very recent given the span of evolutionary time, and that, nevertheless, 5 million years is still a very long time.

A helpful way to think about this is in terms of walking. The human pace is about a yard, so if we call each pace a century, then to walk back to the time of Christ is to walk twenty yards, or two first downs of a football field. With this scale in mind, to walk back to the ape-human divergence is to walk twenty-seven miles, the length of a marathon, which is a long walk indeed until we calculate the distance to the origin of the first animals at the dawn of the Cambrian, 600 million years ago, which is about 3,000 miles or a walk from New York to San Francisco. And to go back to the origin of the Earth, 4.5 billion years ago, is to trek the entire circumference of the planet, 100 years a pace.

PRIMATE SPECIATION

This time digression leads us directly to a consideration of primate evolution. Figure 6 shows two versions of the primate family tree, one constructed fifteen years ago and one current. The differences are obvious. Before 1984 we were able to believe that

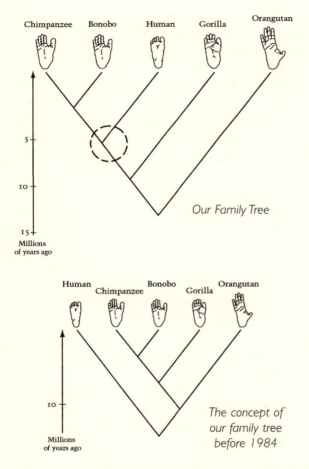

Chimpanzee Bonobo Human Gorilla Orangutan

5

10

15

Millions
of years ago

Our Family Tree

Human Chimpanzee Bonobo Gorilla Orangutan

10

Millions
of years ago

*The concept of
our family tree
before 1984*

∞ FIGURE 6 PRIMATE EVOLUTIONARY TREE ∞ The dotted circle represents the common ancestor to chimpanzees, bonobos, and humans.

we branched off from the other primates some 15 million years ago, and that our evolution was somehow a separate affair, removed from the rest of primate speciation. Now it is clear that we branched off in the thick of things, and only 5 million years ago, with the chimps and bonobos diverging after we did. Importantly, the dotted circle marks an ancestor that was common to chimps, bonobos, and humans. To think about this common ancestor is to absorb important concepts about who we are.

You may recall our example of the ancient Snake-God text, where parts of the text were conserved and parts were reconfigured as it spread into new civilizations. Applying this thinking to the evolution of bonobos, humans, and chimps, we can say that if a trait is evident in all three lineages, it was likely found in the common ancestor as well, meaning that the common ancestor was intelligent and had a chimp body plan, dextrous hands, a complex sociality, robust emotions, and self-recognition in the mirror. We can further say that a trait displayed by only one of the three lineages probably arose after the three-way divergence; this would include symbolic language for humans and sexual co-dominance for bonobos.

And finally, we can say as a first approximation that if a trait shows up in two of the three, it was likely present in the common ancestor as well and subsequently lost by one branch. So, chimps and bonobos are arboreal and hairy, humans are not; therefore, the common ancestor was probably arboreal and hairy and humans lost both traits. Humans and chimps regularly kill animals for food, bonobos do not; therefore, the common ancestor was probably a carnivore and bonobos lost the trait. Female humans and bonobos have concealed ovulation whereas female chimps go into heat; therefore, the common female ancestor probably had concealed ovulation and chimps lost the trait.

Current estimates indicate that human nuclear DNA is 98.4 percent identical to chimpanzee DNA. With our understanding of alleles and speciation, we can say that the alleles in the chimp gene pool will not be identical to those in the human, nor will shared alleles be present with the same frequencies. Still, we struggle with the numbers. There are, after all, lots of differences between us and the apes. How are they accounted for?

The human genome contains approximately 100,000 genes and 1.6 percent of 100,000 is 1,600. Therefore, some of the human-ape differences can readily be accounted for by novel genes. But variation can also arise as the consequence of heterochrony, a phenomenon that can yield major differences between animals without necessitating the evolution of novel genes.

Heterochrony literally means a difference in timing. When we described in an earlier chapter the development of a typical animal embryo we said, as an example, that certain genes were switched on in a particular cell at the eight-cell stage, and this led to a particular cascade of developmental consequences. Were a heterochronic mutation to occur in the promoter of one of these genes, it might not have the correct configuration to switch on until the sixty-four-cell stage, at which point its protein product would encounter a very different spectrum of proteins from those present at the eight-cell stage. Should it form a novel complex with one of these proteins, this might create a new kind of activator that speeds up the timing of the next switching event or perhaps initiates a new switching sequence.

Animal embryology, then, can be thought of in terms of the musical form called theme and variations. The theme evolved prior to the Cambrian explosion. It consists of a set of genes (the Hox cluster) that convey the basic body plan—head, thorax, legs, tail—and a number of additional sets of genes that convey other types of positional information in the embryo (dorsal/ventral, internal/external). Since the Cambrian, this theme has been subjected to all sorts of variations. The conserved sets of genes are expressed at different rates, and in all sorts of temporal and spatial combinations with one another, to produce all manner of embryos, and hence all manner of adults, in the animal radiation.

So, to return to primates. Primate embryology and postnatal development are very complex, with long pregnancies and long periods of brain development. Given that patterns of brain development are highly dependent on which cells are in contact with which, changes in the timing of gene expression during the migration of neurons into the brain could greatly influence the outcome.

In any case, whether by heterochrony or novel genes or both, the chimps, humans, and bonobos emerged as very similar species with some very important differences. In particular, our lineage has come up with the symbol systems that allow both language and self-awareness, apparently the only time in evolutionary history that such capacities have developed. Accompanying this has been a huge amplification in the number of neurons and neuronal connections in the human cerebral cortex and prefrontal cortex, a change evident in the fossil record of skull size about 2 million years ago—about the time we encounter the first tools. The genetic changes that generated our big brains were perhaps not very complicated, but the results are complicated.

Reflections

Distinctiveness. We have encountered this concept several times. We have celebrated our individual selves as organisms, as self-aware creatures, and as recipients of immanence and grace, even as we have also honored the experience of humility, of being but a part of the whole and yet connected to the whole.

Here we can lift up our distinctiveness as a lineage and as a species. Our hominid relatives—the Neanderthals and *Homo erectus*—are no longer with us, but we are privileged to share the planet with our next of kin, the orangs and gorillas and chimps and bonobos. We have much to learn from one another, and the preservation of their habitat and dignity emerges as a commandment.

And then we turn to ourselves. Our gene pool harbors a large number of alleles, a few concentrated in particular geographic groups we call races, which means that each of us is distinctive, and each of our children will be distinctive. But all of us are also members of the human species and hence share the distinctiveness of our species. Indeed, species and special come from the same linguistic root.

As environmentalists we have learned to defend the diversity of species in general, and endangered species in particular. Religious naturalism exhorts us to celebrate human distinctiveness with the same full-throated thanksgiving that we celebrate the whale and the spotted owl. The whale and the owl are magnificent, but so are we.

* We are a symbolic species, unique in our capacity to engage not just in communication but in *language*. As neurobiologist Terrence Deacon puts it, "Biologically, we are just another ape. Mentally, we are a new phylum of organisms." Our symbolic language allows us to build scenarios, plan for the future, and articulate and transmit our cultures. It is the wellspring of our uniqueness.

* While all creatures have the capacity to interpret the reality that they perceive, we also have the capacity to *analyze reality*, to ask questions that yield answers that generate new questions. All of us, that is, are scientists—even as very young children, we construct hypotheses

and test them out. We resonate with the imperative to understand how things work, including our universe and planet and brains and emotions and sexuality and cultures, and we use these understandings to generate our technology and our social systems.

* We have as well the capacity to *take off from reality,* molding it into the distinctively human forms we call art. It is as we respond to the understandings and feelings inherent in our art that we acquire much of our truth, much of our nobility and grace, and much of our pleasure.

* And finally, we are uniquely *religious.* Anthropologists have given the name *Homo religiosus* to our forebears who first buried their dead and set flowers and icons beside the graves. We need answers to existential questions. We need to believe in things, to structure and orient our lives in ways that make sense and offer hope, to identify values and ideals, to transcend and interconnect. And happily, we have the capacity to transmit our accumulated religious understandings to one another and to our children through our languages and our arts, allowing them to endure and evolve.

Emergent Religious Principles

When the responses elicited by the Epic of Evolution are gathered together, several religious principles emerge that can, I believe, serve as a framework for a global Ethos.

TAKING ON ULTIMACY

We are all, each one of us, ordained to live out our lives in the context of ultimate questions, such as:

* Why is there anything at all, rather than nothing?
* Where did the laws of physics come from?
* Why does the universe seem so strange?

My response to such questions has been to articulate a covenant with Mystery. Others, of course, prefer to respond with answers, answers that often include a concept of god. These answers are by definition beliefs since they can neither be proven nor refuted. They may be gleaned from existing faith traditions or from personal search. God may be apprehended as a remote Author without present-day agency, or as an interested Presence with whom one can form a relationship, or as pantheistic—Inherent in All Things.

The opportunity to develop personal beliefs in response to questions of ultimacy, including the active decision to hold no Beliefs at all, is central to the human experience. The important part, I believe, is that the questions be openly encountered. To take the universe on—to ask Why Are Things As They Are?—is to generate the foundation for everything else.

GRATITUDE

Imagine that you and some other humans are in a spaceship, roaming around in the universe, looking for a home. You land on a planet that proves to be ideal in every way. It has deep forests and fleshy fruits and surging oceans and gentle rains and cavorting creatures and dappled sunlight and rich soil. Everything is perfect for human habitation, and everything is astonishingly beautiful.

This is how the religious naturalist thinks of our human advent on Earth. We arrived but a moment ago, and found it to be perfect for us in every way.

And then we came to understand that it is perfect because we arose from it and are a part of it.

Hosannah! Not in the highest, but right here, right now, this.

When such gratitude flows from our beings, it matters little whether we offer it to God, as in this poem, or to Mystery or Coyote or Cosmic Evolution or Mother Earth:

i thank You God for most this amazing
day:for the leaping greenly spirits of trees
and a blue true dream of sky;and for everything
which is natural which is infinite which is yes

(i who have died am alive again today,
and this is the sun's birthday;this is the birth
day of life and of love and wings:and of the gay
great happening illimitably earth)

how should tasting touching hearing seeing
breathing any—lifted from the no
of all nothing—human merely being
doubt unimaginable You?

(now the ears of my ears awake and
now the eyes of my eyes are opened)
                              ✤ *E.E. Cummings, 1950*

REVERENCE

Our story tells us of the sacredness of life, of the astonishing complexity of cells and organisms, of the vast lengths of time it took to generate their splendid diversity, of the enormous improbability that any of it happened at all. Reverence is the religious emotion elicited when we perceive the sacred. We are called to revere the whole enterprise of planetary existence, the whole and all of its myriad parts as they catalyze and secrete and replicate and mutate and evolve.

Ralph Waldo Emerson invites us to express our reverence in the form of prayer. "Prayer," he writes, "is the contemplation of the facts of life from the highest point of view. It is the soliloquy of a beholding and jubilant soul."

CREDO OF CONTINUATION

We have thought of evolution as being about prevalence, about how many copies there are of which kinds of genomes. But it is quite as accurate, and I believe more germinative, to think of evolution as being about the continuation of genomes. Genomes that create organisms with sufficient reproductive success to have viable offspring are able to continue into the future; genomes that fail, fail.

Reproductive success is governed by many

variables, but key adaptations have included the evolution of awareness, valuation, and purpose. In order to continue, genomes must dictate organisms that are aware of their environmental circumstances, evalute these inputs correctly, and respond with intentionality.

And so, I profess my Faith. For me, the existence of all this complexity and awareness and intent and beauty, and my ability to apprehend it, serves as the ultimate meaning and the ultimate value. The continuation of life reaches around, grabs its own tail, and forms a sacred circle that requires no further justification, no Creator, no superordinate meaning of meaning, no purpose other than that the continuation continue until the sun collapses or the final meteor collides. I confess a credo of continuation.

And in so doing, I confess as well a credo of human continuation. We may be the only questioners in the universe, the only ones who have come to understand the astonishing dynamics of cosmic evolution. If we are not, if there are others who Know, it is unlikely that we will ever encounter one another. We are also, whether we like it or not, the dominant species and the stewards of this planet. If we can revere how things are, and can find a way to express gratitude for our existence, then we should be able to figure out, with a great deal of work and good will, how to share the Earth with one another and with other creatures, how to restore and preserve its elegance and grace, and how to commit ourselves to love and joy and laughter and hope.

It goes back in the end to my father's favorite metaphor. "Life is a coral reef. We each leave behind the best, the strongest deposit we can so that the reef can grow. But what's important is the reef."

OUR RELIGIONS OF ORIGIN

So we extract from reality all the meaning and guidance and emotional substance that we can, and we bring these responses with us as we set out to chart global paths. And then we come back to our religions of origin, the faiths of our mothers and fathers. What do we do with them? What have I done with mine?

Theologian Philip Hefner offers us a weaving metaphor. The tapestry maker first strings the warp, long strong fibers anchored firmly to the loom, and then interweaves the weft, the patterns, the color, the art. The Epic of Evolution is our warp, destined to endure, commanding our universal gratitude and reverence and commitment. And then, after that, we are all free to be artists, to render in language and painting and song and dance our ultimate hopes and concerns and understandings of human nature.

Throughout the ages, the weaving of our religious weft has been the province of our prophets and gurus and liturgists and poets. The texts and art and ritual that come to us from these revered ancestors include claims about Nature and Agency that are no longer plausible. They use a different warp. But for me at least, this is just one of those historical facts, something that can be absorbed, appreciated, and then put aside as I encounter the deep wisdom embedded in these traditions and the abundant opportunities that they offer to experience transcendence and clarity.

I love traditional religions. Whenever I wander into distinctive churches or mosques or temples, or visit museums of religious art, or hear performances of sacred music, I am enthralled by the beauty and solemnity and power they offer. Once we have our feelings about Nature in place, then I believe that we can also find important ways to call ourselves Jews, or Muslims, or Taoists, or Hopi, or Hindus, or Christians, or Buddhists. Or some of each. The words in the traditional texts may sound different to us than they did to their authors, but they continue to resonate with our religious selves. We know what they are intended to mean.

Humans need stories—grand, compelling stories—that help to orient us in our lives and in the cosmos. The Epic of Evolution is such a story, beautifully suited to anchor our search for planetary consensus, telling us of our nature, our place, our context. Moreover, responses to this story—what we are calling religious naturalism—can yield deep and abiding spiritual experiences. And then, after that, we need other stories as well, human-centered stories, a mythos that embodies our ideals and our passions. This mythos comes to us, often in experiences called revelation, from the sages and the artists of past and present times.

Notes and Further Reading

ILLUSTRATIONS

The haunting cover photograph was taken by Tui de Roy of New Zealand. It shows the magnificent tortoises *(Testudo nigra)* of Alcedo Volcano, located in the Galapagos archipelago off the coast of Ecuador. Preservation of their habitat is one of the missions of the Charles Darwin Foundation.

The drawings that grace the chapter headings are the work of Ippy Patterson of North Carolina. The organisms illustrated, many of whom inhabit the North Carolina woodlands, are as follows. II *Eoastrion* (fossil cyanobacterial mat). III *Amsinckia grandiflora*. IV *Franklinia alatahaha*. V *Icaronycteris* (fossil bat). VI *Ctenopyge* (fossil trilobytes). VII *Malacothamnus clementinus*. VIII *Callirhoe scabriuscula*. IX *Punica granatum*. X *Oenothera deltoidea* ssp. *Howellii*. XI *Astragalus linifolius*. XII *Convolvulus sepium*, *Rosa fortuniana*, and *Daucus carota*.

INTRODUCTION

My understanding of the nature of religion has been particularly influenced by Loyal Rue; see, for example, *Amythia: Crisis in the Natural History of Western Culture* (Tuscaloosa: University of Alabama Press, 1989), *By the*

Notes and Further Reading

Grace of Guile: The Role of Deception in Natural History and Human Affairs (New York: Oxford University Press, 1994), and his forthcoming book *Everybody's Story*.

Two other key influences have been Erwin R. Goodenough [see Eleanor B. Mattes, *Myth for Moderns: Erwin Ramsdell Goodenough and Religious Studies in America, 1938-1955* (Lanham, MD: Scarecrow Press, 1997) and E. R. Goodenough, "A Historian of Religion Tries to Define Religion," *Zygon* 2: 7–22 (1967)] and William James [*Varieties of Religious Experience: A Study in Human Nature* (New York: Longmans, Green & Co., 1903), a book quoted several times in the text].

The term "Epic of Evolution" is invoked by Edward O. Wilson in *On Human Nature* (Cambridge, MA: Harvard University Press, 1978). The term "religious naturalism" appears in *Science and Religion: A Critical Survey* by Holmes Rolston III (New York: Random House, 1987). Connie Barlow explores the religion/science interface in *Green Space, Green Time: The Way of Science* (New York: Copernicus, 1997).

CHAPTER 1

Several books offer lucid introductions to cosmology and quantum theory: Timothy Ferris, *The Whole Shebang: A State-of-the-Universe(s) Report* (New York: Simon and Schuster, 1997); Steven Weinberg, *Dreams of a Final Theory* (New York: Pantheon, 1992) and *The First Three Minutes: A Modern View of the Origin of the Universe* (New York: Basic Books, 1988); and Stephen Hawking, *A Brief History of Time: From the Big Bang to Black Holes* (New York: Bantam, 1988). A poetic rendition of cosmology is offered by Brian Swimme and Thomas Berry, *The Universe Story: From the Primordial Flaring Forth to the Ecozoic Era—A Celebration of the Unfolding of the Cosmos* (San Francisco:

HarperSanFrancisco, 1992), and by Brian Swimme, *The Hidden Heart of the Cosmos: Humanity and the New Story* (Maryknoll, NY: Orbis Books, 1996).

The quote from Steven Weinberg appears in *The First Three Minutes* (op. cit.).

The hymns quoted in this and subsequent chapters are from *The United Methodist Hymnal* (Nashville, TN: The United Methodist Publishing House, 1989).

Lao Tzu, *Tao Te Ching,* translated by Gia-Fu Feng and Jane English (New York: Vintage Books, 1989).

CHAPTER 11

Four excellent sourcebooks have provided much of my material on comparative religion: John Bowker, *The Oxford Dictionary of World Religions* (New York: Oxford University Press, 1997) and the beautifully illustrated *World Religions: The Great Faiths Explored and Explained* (London: Dorling Kindersley, 1997); Åke Hultkrantz, *Native Religions of North America: The Power of Visions and Fertility* (San Francisco: HarperSanFrancisco, 1987, a volume in the Religious Traditions of the World series); and Dennis Tedlock and Barbara Tedlock, *Teachings from the American Earth: Indian Religion and Philosophy* (New York: Liveright, 1975) from which comes the quote from the Kagaba tribe. Barbara Sproul compiles accounts of creation in *Primal Myths: Creation Myths Around the World* (San Francisco: HarperCollins, 1991).

Current thinking on the origins of life is described in *The RNA World,* Raymond F. Gesteland and John F. Atkins, eds. (Cold Spring Harbor: Cold Spring Harbor Laboratory Press, 1993), *Early Life on Earth,* Stefan Bengston, ed. (New York: Columbia University Press, 1994) and *Comets and the Origin and Evolution of Life,* P. J. Thomas, C. F. Chyba, and C. P. McKay, eds. (New York: Springer, 1997).

Timothy Ferris outlines several versions of the Anthropic Principle in *The Whole Shebang* (op. cit.).

Walt Whitman, "Song of Myself" in *Leaves of Grass*. (New York: Random House, 1921).

CHAPTERS 111 AND 1V

Two excellent accounts of cell and molecular biology are a general-readership book by Boyce Rensberger, *Life Itself: Exploring the Realm of the Living Cell* (New York: Oxford University Press, 1996), and a textbook by Bruce Alberts and colleagues, *Molecular Biology of the Cell*, 3rd ed. (New York: Garland, 1994). Memorable essays are found in Lewis Thomas, *The Lives of a Cell: Notes of a Biology Watcher* (New York: Penguin, 1974).

Embryology is masterfully described in Eric H. Davidson, *Gene Activity in Early Development*, 3rd ed. (Orlando, FL: Academic Press, 1986) and in John Gerhart and Marc Kirschner, *Cells, Embryos, and Evolution* (Cambridge, MA: Blackwell, 1997).

Robert Wright considers reductionism versus holism in his essay "The Great Divide" found in Connie Barlow, ed. *From Gaia to Selfish Genes: Selected Writings in the Life Sciences* (Cambridge, MA: MIT Press, 1991).

CHAPTER V

A key early book on molecular evolution is Jacques Monod, *Chance and Necessity* (New York: Vintage Books, 1972). Of the many excellent books on evolution by Richard Dawkins, I particularly recommend *The Blind Watchmaker* (New York: Norton, 1986). Connie Barlow has edited a fine collection of writings: *Evolution Extended: Biological Debates on the Meaning of Life* (Cambridge, MA: MIT Press, 1994). And Charles Darwin is of course seminal: *Origin of the Species* (London: John Murray, 1859; reprinted, New York: Penguin Classics, 1984).

Mary Oliver, "Wild Geese," in *Dream Work* (New York: Atlantic Monthly Press, 1986).

CHAPTER VI

Edward O. Wilson's *The Diversity of Life* (Cambridge, MA: Harvard University Press, 1992) gives an excellent introduction to the topic, and a fine book on the relationships of the major phyla is Peter J. Bowler, *Life's Splendid Drama: Evolutionary Biology and Reconstruction of Life's Ancestry* (Chicago: University of Chicago Press, 1996). Beautiful contemporary essays on Nature come from Aldo Leopold, *A Sand County Almanac* (New York: Oxford University Press, 1949), Annie Dillard, *Pilgrim at Tinker Creek* (New York: Harper and Row, 1974) and Chet Raymo, *Honey From Stone: A Naturalist's Search for God* (St. Paul: Hungry Mind Press, 1987); classic essays come from Henry David Thoreau ("Walden") and Ralph Waldo Emerson ("On Nature").

J. Baird Callicott analyzes religions from an environmental perspective in *Earth's Insights: A Multicultural Survey of Ecological Ethics from the Mediterranean Basin to the Australian Outback* (Berkeley: University of California Press, 1994), as do Mary Evelyn Tucker and John Grim in *Worldviews and Ecology* (Maryknoll, NY: Orbis Books, 1994).

The quote from Oren Lyons is found in *The Essential Mystics: Selections from the World's Great Wisdom Traditions,* Andrew Harvey, ed. (San Francisco: HarperSanFrancisco, 1996). The Pawnee prayer is found in a particularly rich collection of Earth-centered poetry called *Earth Prayers from Around the World: 365 Prayers, Poems, and Invocations for Honoring the Earth* (San Francisco: HarperSanFrancisco, 1991), edited by Elizabeth Roberts and Elias Amidon, and originally appeared in *The Hako, A Pawnee Ceremony,* © Smithsonian Institute, Washington,

D.C., 1904. Figure 3 was provided by Mitchell L. Sogin.

CHAPTERS VII AND VIII

Three excellent books provide details on the neurobiology summarized in these chapters: Antonio R. Damasio, *Descartes' Error: Emotion, Reason, and the Human Brain* (New York: Putnam, 1994) (the quote in Chapter VIII is found on p. xvi); Joseph LeDoux, *The Emotional Brain: The Mysterious Underpinnings of Emotional Life* (New York: Simon and Schuster, 1996), and Steven Pinker, *How the Mind Works* (New York: Norton, 1997).

Robert Wright considers our psychological evolution in *The Moral Animal: Why We Are the Way We Are: The New Science of Evolutionary Psychology* (New York: Pantheon, 1994).

The quote from Albert Einstein comes from *The World As I See It* (New York: Philosophical Library, 1934).

Figure 4 derives from W. Hudos, "Evolutionary interpretation of neural and behavioral studies of living vertebrates," in F. O. Schmidt, ed., Neurosciences: Second Study Program, 1970, page 31, by copyright permission of the Rockefeller University Press.

"Spirit of the Living God," Daniel Iverson, © 1935, 1963 Birdwing Music (ASCAP), administered by EMI Christian Music Publishing. All rights Reserved. Used by Permission.

CHAPTERS IX AND X

George C. Williams has written a classic book, *Sex and Evolution* (Princeton, NJ: Princeton University Press, 1975) and more recent reviews are found in *The Evolution of Sex: An Examination of Current Ideas* (Sunderland, MA: Sinauer, 1988), edited by Richard Michod and Bruce Levin. An evolutionary perspective on human sexuality is offered by Helen Fisher in *Anatomy of Love: A Natural History of*

Mating, Marriage, and Why We Stray (New York: Fawcett Columbine, 1992).

"High School Senior," from *The Wellspring* by Sharon Olds, © 1996 by Sharon Olds, reprinted by permission of Alfred A. Knopf, Inc.

Michelangelo Buonarroti (1475–1564), "Veggio co' be' vostr' occhi," 1564. Sonnet for Tommaso de' Cavalieri from *Set My Heart Aright: A Michelangelo Portrait*, translations and music © Carl F. Smith, 1996.

CHAPTER XI

A collection of essays on death, edited by J. D. Roslansky, is found in *The End of Life* (Amsterdam: North-Holland Publishing, 1973), one of which, "The Origin of Death" by George Wald, also considers the importance of the germ/soma dichotomy. William R. Clark has developed this idea as well in *Sex and the Origins of Death* (New York: Oxford University Press, 1996).

CHAPTER XII

A classic and still durable textbook on speciation is Steven Stanley, *Macroevolution: Pattern and Process* (San Francisco: Freeman, 1979), from which Figure 5 derives. Human evolution has been considered in a number of excellent books, including: Merlin Donald, *Origins of Modern Mind: Three Stages in the Evolution of Culture and Cognition* (Cambridge, MA: Harvard University Press, 1991); Richard Wrangham and Dale Peterson, *Demonic Males: Apes and the Origins of Human Violence* (Boston: Houghton Mifflin, 1996), from which Figure 6 derives; Frans de Waal and Frans Lanting, *Bonobo: The Forgotten Ape* (Berkeley: University of California Press, 1997); and Terrence W. Deacon, *The Symbolic Species: The Co-Evolution of Language and the Brain* (New York: Norton, 1997), whose quote appears on p. 23.

The walking/time analogy is developed by Richard Dawkins in *River Out of Eden: A Darwinian View of Life* (New York: Basic Books, 1995).

EMERGENT RELIGIOUS PRINCIPLES
I am grateful to Richard Dawkins for narrating the spaceship metaphor to me; his wonderful version is found in *Unweaving the Rainbow* (New York: Houghton Mifflin, 1998).

E.E. Cummings, "i thank you God" in: R.S. Kennedy, ed., *Selected Poems of E.E. Cummings* (New York: Liverright, 1994).

Ralph Waldo Emerson, "Self-Reliance" in: R. Poirier, ed., *Ralph Waldo Emerson* (Oxford: Oxford University Press, 1990).

Philip Hefner, "The Spiritual Task of Religion in Culture," *Zygon 32*: (in press), 1998.

Loyal Rue considers religious myth in "Redefining Myth and Religion: Introduction to a Conversation" *Zygon: 29*: 315–19, 1994. A classic book on myth is Joseph Campbell, *The Masks of God: Primitive Mythology* (New York: Viking, 1959).

Index

Weaving metaphor, 172-173
Web of life, 64, 87
Weinberg, Steven, 10
Wesley, Charles, 138
Whale, 165
Whitman, Walt, 30-31
Working memory. *See* Memory

Worm, 97

Yaruro, 17
Yearning, 30, 47, 59, 60, 137-141
Yeast, 72, 111, 132

Zygote, 119, 121, 122, 143, 144, 147, 153-154